Consolationscapes in the Face of Loss

T0199714

Human beings are grieving animals. 'Consolation', or an attempt to assuage grief, is an age-old response to loss which has various expressions in different cultural contexts. Over the past century, consolation has dropped off the West's cultural radar. The contributions to this volume highlight this neglect of consolation in popular and academic discourses and explore the usefulness of the concept of consolation for analysing spatio-temporal constellations.

Consolationscapes in the Face of Loss brings together scholars from geography, philosophy, history, anthropology and religious studies. The chapters use spatial and conceptual mappings of grief and consolation to analyse a range of spaces and phenomena around grief, bereavement and remembrance, comfort and resilience, including battlefield memorials, crematoria, graveyards and natural burial sites in Europe. Authors shift the discussion beyond the Global North by including responses to traumatic grief in post-conflict African societies, as well as Australian Aboriginal traditions of ritual consolation.

The book focuses on the relationship between space/place and consolation. In so doing, it offers a new lens for research on death, grief and bereavement. It offers new insights for students and researchers interrogating contemporary bereavement, as well as those interested in meaning-making, emerging socio-cultural practices and their role in personal and collective resilience.

Christoph Jedan is Professor of Ethics and Philosophy of Religion at the University of Groningen, the Netherlands.

Avril Maddrell is Professor of Social and Cultural Geography at the University of Reading, UK.

Eric Venbrux is Professor of Comparative Religion and Director of the Centre for Thanatology at Radboud University, the Netherlands.

Routledge Studies in Human Geography

This series provides a forum for innovative, vibrant, and critical debate within Human Geography. Titles will reflect the wealth of research which is taking place in this diverse and ever-expanding field. Contributions will be drawn from the main sub-disciplines and from innovative areas of work which have no particular sub-disciplinary allegiances.

Living with the Sea
Knowledge, Awareness and Action
Edited by Mike Brown and Kimberley Peters

Time Geography in the Global Context
An Anthology
Edited by Kajsa Ellegård

Consolationscapes in the Face of Loss
Grief and Consolation in Space and Time
Edited by Christoph Jedan, Avril Maddrell and Eric Venbrux

The Crisis of Global Youth Unemployment
Edited by Tamar Mayer, Sujata Moorti and Jamie K. McCallum

Time Geography
Kajsa Ellegård

British Migration
Globalisation, Transnational Identities and Multiculturalism
Edited by Katie Walsh and Pauline Leonard

For more information about this series, please visit: www.routledge.com/Routledge-Studies-in-Human-Geography/book-series/SE0514

Consolationscapes in the Face of Loss

Grief and Consolation in Space and Time

Edited by
Christoph Jedan, Avril Maddrell and
Eric Venbrux

Routledge
Taylor & Francis Group

LONDON AND NEW YORK

First published 2019
by Routledge
2 Park Square, Milton Park, Abingdon, Oxon OX14 4RN

and by Routledge
52 Vanderbilt Avenue, New York, NY 10017

First issued in paperback 2020

Routledge is an imprint of the Taylor & Francis Group, an informa business

British Library Cataloguing-in-Publication Data
A catalogue record for this book is available from the British Library

Library of Congress Cataloging-in-Publication Data
A catalog record has been requested for this book

ISBN 13: 978-0-367-58431-3 (pbk)
ISBN 13: 978-0-8153-5879-4 (hbk)

Typeset in Times New Roman
by Taylor & Francis Books

To Albertina (Tineke) Nugteren

Contents

Illustrations

Figures

Tables

Contributors

Sophie Bowlby is Visiting Professor, University of Loughborough, and Visiting Research Fellow, Department of Geography and Environmental Science, University of Reading, UK. She was a consultant on the 'Death in the Family in Urban Senegal' research project. Her research has focused on feminist analysis of the social and economic geography of urban areas in the UK, in particular, issues of access, mobility and the analysis of social relationships of informal care in time-space.

Ruth Evans is Associate Professor in Human Geography, Department of Geography and Environmental Science, University of Reading, UK. She was the Principal Investigator of the research project, 'Death in the Family in Urban Senegal: Bereavement, Care and Family Relations', funded by The Leverhulme Trust (2014–2016). Her research interests focus on young people's psychosocial wellbeing, care and family relations in relation to bereavement, chronic illness and forced migration. See https://blogs.reading.ac.uk/deathinthefamilyinsenegal/.

Martin J. M. Hoondert studied musicology and theology and specialises in music and rituals. Since 2007 he has been (Associate) Professor of 'Music, Religion & Ritual' at the Department of Culture Studies of Tilburg University, the Netherlands. His research focuses on 'music and death' and 'practices of memorialisation'. His research topics are: the contemporary Requiem, musical repertories of funeral rites, commemoration and music, music and grief, music and the First and Second World War, practices of memorialisation regarding genocide (esp. Rwanda and Srebrenica). In cooperation with the universities of Bath (UK), Leuven (Belgium) and San Sebastian (Spain) he has established a network focusing on research into practices of memorialisation and the process of social reconstruction after atrocities. Related to this research network he co-edited the book *Cultural Practices of Victimhood* (2018).

Christoph Jedan is Professor of Ethics and Philosophy of Religion in the Faculty of Theology and Religious Studies, University of Groningen, the Netherlands. His research interests include Ancient Greek and Roman philosophy, the intersections of religion and philosophy today (religion and politics, postsecularism), and the history and continuing relevance of consolation for death and loss. Among his earlier books is *Stoic Virtues: Chrysippus and the Religious Character of Stoic Ethics* (2009). He has edited and co-edited half a dozen books and special journal issues, among them *Exploring the Postsecular: The Religious, the Political and the Urban* (2010).

Fatou Kébé is Researcher, Maternal and Child Health Intervention, International Development Research Centre (CRDI), in collaboration with ACDEV, Ministry of Health and Université Cheikh Anta Diop de Dakar (UCAD), Senegal. She worked as Researcher on the 'Death in the Family in Urban Senegal' research project, based at the Laboratoire de Recherches sur les Transformations Economiques and Sociales, Institut Fondamental d'Afrique Noire, UCAD. Her research interests focus on street children, poverty, education, health and migration in Senegal.

Anne Kjærsgaard is Postdoctoral Researcher in the Department of Practical Theology and Church History, School of Society and Culture, Aarhus University, Denmark. Her research interests concern lived religion and material culture in relation to mortuary practices. Kjærsgaard conducted her doctoral research at the Centre for Thanatology, Radboud University, the Netherlands. She is the author of *Funerary Culture and the Limits of Secularization in Denmark* (2017) and several book contributions and journal articles in the field of death studies. She is one of the editors of the Danish journal for churchyard culture, *Kirkegårdskultur*.

Jane Ribbens McCarthy is Reader in Family Studies (retired) and Visiting Fellow, Open University, and Visiting Professor, University of Reading, UK. She was co-investigator for the 'Death in the Family in Urban Senegal' research project. Her research interests focus on people's family lives and relationships, drawing on a framework of 'family troubles', as these are shaped across diverse global and local contexts. See http://www.open.ac.uk/people/jcrm2.

Dolly MacKinnon is Associate Professor of Early Modern History at the University of Queensland, Australia. She has an interdisciplinary background with a Bachelor in Music and a PhD in early modern history, both from the University of Melbourne. Her research analyses the mental, physical and auditory landscapes of past cultures.

Avril Maddrell is Professor of Social and Cultural Geography at the University of Reading, UK, and co-editor of the journal *Social and Cultural Geography*. She has authored, co-authored and co-edited eight books, including *Deathscapes: Spaces for Death, Dying, Mourning and Remembrance* (2010), *Memory, Mourning, Landscape* (2010), *Christian Pilgrimage, Landscape and Heritage* (2015) and *Sacred Mobilities* (2015).

Albertina Nugteren has an academic background in South Asian languages and cultures. She works as a Religion-and-Ritual specialist at the Faculty of Humanities, Tilburg University, the Netherlands. Current research topics include: (1) the nexus Nature–Culture–Religion (recent example: 'Sacred Trees, Groves and Forests' in *Oxford Bibliographies in Hinduism* [2018]); (2) Funerary rituals, particularly environmental aspects (recent example: 'Wood, Water and Waste: Material Aspects of Mortuary Practices in South Asia' [2017]); (3) Critical discourses on the 'greening of religion' (recent example: 'A Darker Shade of Green? An Inquiry into Growing Preferences for Natural Burial' [2015]); (4) object-centred studies of ritual (recent example: guest editorship of a Special Issue on 'Religion, Ritual and Ritualistic Objects' in *Religions* [2018]).

Sophie Seebach is the curator of the Ethnographic Collections at Moesgaard Museum in Aarhus, Denmark. She holds a PhD in anthropology from Aarhus University, and has conducted her fieldwork in Gulu, northern Uganda. The title of her PhD dissertation is *The Dead are not Dead: Intimate Governance of Transitions in Acholi* (2016). The focus of her research is death, dying, and burial rites, and how the practices surrounding death are being affected by social change. She is the co-editor of *Mirrors of Passing: Unlocking the Mysteries of Death, Materiality, and Time* (2018).

Joram Tarusarira is Assistant Professor of Religion, Conflict and Peacebuilding at the University of Groningen, the Netherlands. His research interests include religion and conflict transformation, peacebuilding and reconciliation. He is the author of *Reconciliation and Religio-political Non-conformism in Zimbabwe* (2016) and several book chapters and journal articles at the intersection of religion, politics, conflict transformation and reconciliation.

Eric Venbrux is Professor of Comparative Religion and Director of the Centre for Thanatology at Radboud University, the Netherlands. He conducted anthropological fieldwork among the Tiwi people of Australia, as well as in Switzerland and the Netherlands. His research interests are local religion, ritual change, material culture, and the verbal and visual arts. He is author of *A Death in the Tiwi Islands: Conflict, Ritual and Social Life in an Australian Aboriginal Community* (1995), co-editor of *Exploring World Art* (2006), *Ritual, Media, and Conflict* (2011) and *Changing European Death Ways* (2013), and has written numerous articles and chapters on mortuary ritual.

Joséphine Wouango is Lecturer, University of Liège, Belgium, and Research Consultant, female genital mutilation/cutting, Population Council, Kenya. She worked as Research Fellow on the 'Death in the Family in Urban Senegal' research project, based in the Department of Geography and Environmental Science, University of Reading, UK. Her research interests focus on public policies on child labour, social protection, education, children's rights, gender equality and women's rights.

Foreword

Grieving accompanies numerous events in our lives, usually the separation from a loved one such as through divorce and death. We suffer mentally and emotionally, sometimes even physiologically, during a period of grieving. Grieving may be experienced collectively, but perhaps more often, endured alone. Grieving has both a public dimension and an intensely private one. Public displays of grief sometimes reflect the status of the deceased (as when professional mourners are hired) and reflect the values privileged by society.

Grief may be ameliorated through consolation, the effort made to comfort someone who faces separation and loss. We console others in numerous ways, whether it is in expressing regret for the loss, voicing empathy, or in sharing insights for hope for the future. Consolation involves rendering available coping strategies for those experiencing loss and grief, and thereby, hopefully building resilience to face the present and the future for those grieving. Just as grief may be intensely felt in certain spaces and places associated with loss, such as in a hospital, a grave site or at a memorial, so too may consolation be closely linked to space and place, indeed, sometimes the same spaces and places elicit grief and consolation simultaneously. Further, for some, a place associated with grief may for others be a place of solace, comfort and consolation, such as a memorial site.

It is precisely the concept, sites and practices of consolation that this volume is dedicated to analysing. Recognising the paucity of scholarly literature on the topic, the editors have brought together chapters that examine *consolationscapes* in the European and Global South contexts, focusing in particular on consolationscapes associated with death. It is a natural companion volume to the earlier book *Deathscapes: Spaces for Death, Dying, Mourning and Remembrance* (2010), edited by Avril Maddrell and James Sidaway. The multidisciplinary lenses brought to bear on the topic bring the reader through different historical eras, emphasising the *longue durée* of ideas about consolation, as well as the emotional and affective geographies of loss, grief and remembrance and the associated therapeutic spatialities. These temporal and spatial ranges associated with consolation and its related 'scapes' addressed in the first section of the book set the context for the subsequent sections that are organised by geographies, focusing first on Europe and latterly on the Global South.

The volume truly comes alive (no pun intended!) as the reader encounters, *inter alia*, music theory to shed light on the role of music in ritual, ecology to lend perspective to environmental concerns in natural burial, ritual studies to understand the role of drama in consolation, and Freudianism to understand the uncanny and closure. Through the chapters, we are reminded that, in as much as death foregrounds the most important social and cultural values that we live our lives by, so too do the spaces and practices of consolation serve as medium and outcome of those values that we acknowledge and express, as well as those that are neither ordinarily recognised nor explicit.

The chapters remind us that we are assisted in the management of grief associated with death through practices of consolation, including marking and memorialisation, which may vary across cultures. Yet, the fundamental need for consolation, and the relayance of that consolation through spaces and places bears remarkable constancy across cultures and scales – through the body, the home, the sites of bereavement and grief, the spaces of faith, and increasingly, virtual online sites.

Reading the chapters reminds me that some fundamental human needs are thus addressed through such consolationscapes. Understanding how consolationscapes work in relation to death and dying, loss and grieving, offers insights into the needs of those alive, and into the power of place. It is a reminder that our encounters with space and place are fundamentally anchored in emotions, not merely functions. The volume covers much ground, and offers new insights into how, as human beings, we cope with some of the most intensely emotional experiences in our lives. In bringing together multi-disciplinary perspectives to bear on a common human experience, the editors and authors help to revive the study of consolation from relative neglect.

Lily Kong
Lee Kong Chian Chair Professor of Social Sciences
Singapore Management University

Preface

The idea for the present volume was conceived during the 'Emotional Geographies' conference held in Groningen in 2013. Christoph Jedan and Eric Venbrux had proposed a session entitled 'Consolation-scapes: Analysing grief and consolation between space and culture'. Colleagues' interest was phenomenal. Instead of one session, as originally planned, we were able to schedule a series of four. Avril Maddrell not only presented a much-needed theoretical framework but she was immediately prepared to jointly edit a volume that would become in many ways a sequel to *Deathscapes*, the collection that she and James Sidaway had published in 2010. As always happens, busy academic schedules stood in the way of speedy completion. In this case, we think that the book has benefitted from the delay. The chapters that our readers now see are not brushed-up conference papers; they are completely re-thought perspectives on the intertwining of consolation and space in time. Moreover, by including authors who were not present at the Groningen conference, we have been able to broaden the discussion beyond the Global North.

In the long process of preparing this book, we have incurred many debts of gratitude. We thank two anonymous reviewers at Routledge for their enthusiastic support and helpful suggestions, and Michael Rice for copy-editing seven of the chapters. Probably we were a very headstrong and demanding bunch of editors – a good reason to thank Ruth Anderson at Routledge for her kind, patient as well as efficient support.

We are most grateful to Lily Kong, who was so kind as to write a Foreword for the volume, just as she did for *Deathscapes*, edited by Avril Maddrell and James Sidaway.

As editors, we would like to express our gratitude to all of our contributors for their inspiring ideas, and also for the images they have provided. With one exception, all photographic work is theirs. The exception is the photograph of the Grave of the Unknown Warrior (Figure 1.2) in Westminster Abbey. For permission to reproduce it, we are grateful to the Dean and Chapter.

We would like to specifically highlight Tineke Nugteren's contribution. Although she was stricken by a life-threatening illness, she crafted a chapter that is a model of interdisciplinary engagement and a testimony of resilience in the face of adversity. It is for this reason that we dedicate *Consolationscapes in the Face of Loss* to her.

Christoph Jedan, Avril Maddrell and Eric Venbrux
Groningen – Reading – Nijmegen

Introduction

From deathscapes to consolationscapes: spaces, practices and experiences of consolation

Christoph Jedan, Avril Maddrell and Eric Venbrux

Human beings are grieving animals and consolation, an experiential assemblage through which grief is ameliorated or assuaged, is an age-old response to loss, expressed variously in different cultural contexts. However, in the context of the West, over the course of the past century, consolation has dropped off the cultural radar, reduced in popular usage to the notion of 'second prize' rather than any positive agential process.[1] It might seem that we don't 'do' consolation any more, and Western models of bereavement in the twentieth century typically privileged coping with loss as a linear progression towards 'closure'. The contributions to this volume highlight this relative neglect of consolation in Western popular and academic discourses *and* show that the international traditions of consolation discussed here illuminate diverse attitudes to death and offer insight to a range of strategies for dealing with bereavement across different cultures, and the varied ways in which grief and consolation are intertwined with the spatial fabric of social worlds across different cultural settings. Indeed, in the context of this volume on *consolationscapes*, the 'scapes' suffix is important: 'scapes' are concerned with contact zones, spaces of exchange, nexus, i.e. multidimensional. Thus, this volume moves beyond consideration of single thematic approaches to consolation, such as religion, to explore the ways in which varied relational factors coalesce, are expressed, mapped on to, and experienced in and through particular places, including the body, home, landscapes and the virtual arenas of online sites and communities and faithscapes.

Bringing a spatial lens to consolation

Bereavement results in perceiving, inhabiting and experiencing some spaces, places, material and immaterial arena in new and different ways, e.g., through bereavement-induced (im)mobilities, or as sites of comfort and ongoing attachment (i.e. as actual or potential sites of consolation). Further work is

---

1   Cf. the *Concise Oxford Dictionary* (Sykes 1982: 201): 'Consolation n. act of consoling; consoling circumstance; - prize (given to competitor just missing main prizes)'.

needed in order to begin to understand the detailed interplay of complex intersectionalities in the time-spaces of death, bereavement and living with loss at individual and communal levels in different regional and national contexts, as well as in different types of social spaces. This includes consolation, which, with notable exceptions (e.g. Maddrell 2009a), has tended to be implicit rather than explicit in work on geographies and other spatially inflected studies of dying, death, loss and mourning.

This book goes some way to promoting this dialogue through its focus on the relationship between space-place and consolation. In so doing, the book offers a fresh perspective for research on death, grief and bereavement, notably in offering a counterpoint to the *Deathscapes* edited collection (Maddrell and Sidaway 2010) (although aspects of comfort, support and consolation featured in several of the chapters found there, e.g. see Watts [2010] on self-help groups).

Consolationscapes in the Face of Loss offers a multidisciplinary collection of spatially attentive studies, which offer new insights for researchers and students interrogating contemporary bereavement and living with loss within geographies of death and wider interdisciplinary death studies, as well as those interested in emerging social-cultural practices, meaning-making and their role in personal and collective resilience. Through the conceptual discussions and case studies which follow, we argue that understanding consolation and its spatial expression offers key insights central to the coping strategies and resilience of those who experience bereavement and that this, in turn, can provide insight to other forms of loss and associated responses and strategies. In turn, we hope that this volume will speak to emerging scholarship on geographies of resilience (Weichselgartner and Kelman 2015) and solace (*Geographies of Comfort*, Price et al. 2018).

The volume is largely based on presentations given at sessions at the Emotional Geographies conference in Groningen in 2013, convened on the theme of 'Consolationscapes'. With some additional contributions, it brings together scholars from geography, sociology, anthropology, history, philosophy and religious studies. The chapters use various spatial and conceptual mappings of grief and consolation to analyse a range of experiential arenas associated with grief, bereavement and remembrance, comfort and resilience, including battlefield memorials, crematoria, maternity wards, graveyards and burial sites, as well as the related everyday ritual performances and practices marking loss and consolation, and how these enable those living with loss to carry on.

Interdisciplinary approaches to consolation

We have claimed above that there has been a relative neglect of consolation in post-war scholarship. The word 'consolation' is not absent from recent scholarly literature, but has been addressed only intermittently and unsystematically. What is evident is the use of 'consolation', almost interchangeably with words such as 'comfort', 'coping', 'resilience' and 'solace', as an indication of an unspecific amelioration of the bereaved's experience and situation (e.g. Kolcaba

2003; Gillis 2012). The wars of 1914–18 and 1939–45 and the associated large-scale deaths of militia and civilians may have marginalised discourses of consolation in Europe in post-war years. Likewise, the notion of consolation may seem hollow in the face of mass disaster, or genocide, such as that of European Jews, as well as Romanies, homosexuals and the disabled, under the Nazi regime, and that of Pol Pot's Khmer Rouge in Cambodia. Trends to secularisation in the West have also been associated with shifts in discourses of consolation (see below); and some memorial culture suggests a passive inclination for forgetting rather than dwelling on loss (Robinson 2010).

There have been a few attempts at offering a conceptual framework of consolation (e.g. Weyhofen 1983; Norberg, Bergsten and Lundman 2001; Klass 2013, 2014), and these represent much needed interventions given the polyvalence of consolation. A good indicator of the complexities is provided by dictionary definitions of *consolatio*, the Latin word from which the English 'consolation' is derived. The *Oxford Latin Dictionary* (Glare 1996) distinguishes three main meanings, first 'the act of consoling or an instance of it, consolation; the title of books, e.g. by Cicero and Crantor', which the dictionary combines with the meaning of 'the act of allaying (fears)'; second, 'the fact of being consoled' and, third, 'a consoling fact or circumstance'. It is notable that the first meaning pulls together the act of consolation and a specific historical genre, written texts intended to offer consolation. To pursue the etymology further, the verb *solari*, from which the noun *consolatio* is derived, combines the meanings of assuaging grief and 'relieving from physical pain or discomfort'. These dictionary definitions thus point us to a number of complex conceptual issues:

1   What is the relationship between consolation and historically situated human experience?
2   To what extent is historical material helpful in understanding consolation today?
3   How do the spatial and temporal aspects of consolation intersect?
4   What is the relationship between activity and passivity; can consolation be actively sought?
5   Is there a relationship between offering and receiving or finding consolation?
6   How do we apprehend and understand the embodied dialectic between grief and consolation?
7   Where is consolation situated within the affective-emotional arena and how does it relate to other affective-emotional registers?

In the light of these challenges we acknowledge the value of preceding work and the limits of this single volume, as well as the aspiration to stimulate further work in this area.

In the humanities – history, classical languages, philosophy, theology and parts of religious studies – research on consolation has always retained a limited presence. This residual presence can be attributed to those disciplines which address historical material in which the concept of consolation is employed. Within historical studies, ancient consolations have attracted detailed studies (e.g. Kassel 1958; Johann 1968, Manning 1974; Alonso del Real 2001; Lillo Redonet 2001; Baltussen 2013; Jedan 2014a, 2014b on the philosophical material; on early Christian consolations e.g. Favez 1937, Scourfield 1993; Holloway 2001; Jedan 2017), as have later periods (e.g. Moos 1971–1972 on the medieval material; Rittgers 2012 on the late medieval and early modern periods; McClure 1991 on Italian Humanism; Resch 2006 on the Protestant Reformation; Simonds and Rothman 1992 on pregnancy loss since the nineteenth century; also relevant are Laqueur 2015 on the birth of the modern cemetery; Roth 2002, Davies 2002 and Bregman 2012 on sermons and funerary rites; Gilbert 2006 on death culture through the lens of literary criticism; and, finally, Kjærsgaard 2017 on death culture in present-day Denmark). There have been relatively few longitudinal studies of consolation as idea and set of practices, and those that are available indicate decline over time (e.g. Weyhofen 1983, with a phenomenal range). In turn these have influenced the field of pastoral theology, traditionally a stronghold of engagement with consolation, whereby contemporary pastoral theology tends to replicate the broader cultural suspicion of consolation *per se* (e.g. Langenhorst 2000).

However, in recent decades there have been attempts to engage with and recover consolation as a concept and analytical tool – of which this volume is part. In the interdisciplinary field of death studies, for instance, questions of solace, comfort, coping, resilience and consolation in bereavement have been both implicit and explicit (see, for example, Klass et al. 1996, Goss and Klass 2005, and Klass and Steffen 2017 on the common experience of continuing bonds with the deceased). Rugg (2018) has recently explored the individuation of funerary rituals in the 'West' as a source of consolation and the threat to this from increasingly impersonal industrialised funerary practices. Klass' (2014) framework for the consolatory function of religion is of particular significance here, which highlights the importance of 'cultural/religious resources that range from the literal image of God as an idealized parent to the abstract architecture of Brahm's Requiem' (Klass 2014: 1). Also see Jedan and Venbrux (2014), a collection of papers noteworthy for its attempt to trace consolation from ancient to contemporary material (music, newspaper columns, cemeteries, modern theology).

Anthropological studies of lament, wailing and related practices highlight aspects of consolation, for example, sharing the pain of the loss or providing 'words against death' (Davies 2002). The spatial dimension of consolatory practices can be seen in the economy of exchanges within the house of mourning among Yemenite Jews in Israel (Gamliel 2014) or spatial metaphors within rural Greek women's laments, which as Danforth (1982: 142) explains, help to create 'a feeling of comfort and solace' in pretending a return of the departed would be possible. Although not discussed explicitly,

Seremetakis' (1991) ethnography of Inner Mani, Greece, cleverly analyses the use of space in mortuary behaviour, including forms of 'spatial intimacy' (Seremetakis 1991: 96). Lotte Buch Segal's study *No Place for Grief* (2016) makes clear that religious and political frames in occupied Palestine leave no further space than acceptance of the consolation of martyrdom by the bereaved mothers in question. In more everyday contexts in rural Europe before the 1960s, Roman Catholic mothers bereft of a small child used to be congratulated for having a little angel in the other world (Venbrux 1991: 197; Behar 1991: 354). The sense of their child being in heaven and the benefit of having a personal mediator in heaven was supposed to offer solace and bring consolation.

Anthropological theory suggests that people turn to religion, when they are confronted with chaos – unable to comprehend, facing injustice or prone to suffering – to cope with such circumstances (Geertz 1973: 100). Losing someone close might be overwhelming on all these counts (intellectually, morally and emotionally), especially in the case of an untimely death. For Klass (2014: 11) religious solace is 'woven into the connection with transcendent reality, the worldview of religious narratives, and the religious community with which people identify'. 'Solace alleviates, but not removes, sorrow or distress', according to Klass (2013: 609). Klass (2014) considers continuing bonds with the dead, often supported by religion, particularly consoling (see also Goss and Klass 2005). However, attention to ancestor-related religious beliefs and practices underscores that consolation should not only be seen as a matter for the living (e.g. Kong 1999). Kwon's ethnography (2013) of spirit consolation rites in Vietnam makes clear that the dead who were victims of atrocities might also be in need of solace themselves.

Within geographical work, studies have highlighted changing life expectancy and consequent attitudes to mortality and consolation in different periods of history, e.g. *Children Remembered* (Woods 2006) and religious practices (Kong 1999). Theoretically, contemporary geographical considerations of consolation sit at the interface of the more-than-representational geographies of religion and spirituality, and of emotion and affect, as well as material and embodied geographies (see Bondi et al. 2005). Building on discussions of continuing bonds and the deceased as a negotiation of relational absence-presence, Maddrell has highlighted the consolation found in this sense of the dynamic absence-presence of the deceased which can be manifest in individual and collective emotional-affective geographies (2012, 2013). Likewise, material and representational vernacular memorials can act as focal points mediating loss and consolatory performative practices (2009a, b). Situating both bodily remains and memories in 'ideal' locations e.g. places associated with memories and identity formation, such as faithscapes and/or natural burial grounds (Maddrell 2011) has also been identified with comfort and the assuagement of grief, as have the virtual spaces of family relations, hospitality and funerary rites in various cultural contexts (Dunn 2016; Evans et al. this volume; Maddrell 2016). More instrumentalist activities such as fundraising and lobbying for causes associated with the deceased and/or the cause of their death can also offer some consolation. These processes are notable in the case of unpredicted, untimely and avoidable death, whereby the sense of consolation grows from ensuring changes take place – in medical research, in safety systems, in infrastructure – in order to reduce the likelihood of the tragedy recurring, being inflicted upon others (Maddrell 2013), processes that can morph into political campaigns (Stevenson et al. 2016).

The following chapters build on and contribute to this body of historical and emerging contemporary engagements with varying ideas of consolation and diverse place-temporalities of consolatory practices.

The chapters

Following this Introduction, the chapters are organised into three broad sections. The first section explicitly addresses consolation as a concept, its dialogic and dialectical relationship to grief, including the links between the *Consolationscapes in the Face of Loss* and *Deathscapes* volumes. The second section examines sites and practices of consolation in the European context, and the third section enriches our understanding of consolation conceptually through attention to case-studies drawn from the Global South.

In the first section, Christoph Jedan and Avril Maddrell explain what consolation is, how it functions and how it is spatially constituted. Both chapters thus respond to the scholarly challenge we are facing today: 'We have (...) extremely limited scholarly or pastoral notions of what consolation is' (Klass, unpublished). Christoph Jedan's chapter rests on the premise that previous models of consolation have been too limited, failing to extrapolate a comprehensive conceptual framework from the wealth of historical material, such that they are either strongly social-scientific models with little connection to a broader range of material, or they are conceptions based on humanities approaches that fail to connect to social scientific expertise. The new 'Four-Axis Model' is designed to capture abiding consolatory themes against a background of historical change, from Greco-Roman Antiquity to today's world of internet memorials and bereavement psychology. The model allows us to compare changing emphases in consolation. Jedan argues further that the historical-conceptual perspective helps in solving an important conundrum: on the one hand, the concept of consolation appears to offer an astonishingly useful analytical tool for analysing spaces and practices to do with loss and bereavement, yet at the same time the word 'consolation' is greeted with a certain reservation. In fact, as long-term linguistic usage can show, we tend to avoid the word 'consolation' in favour of other concepts, such as 'coping'. Jedan's historical and conceptual analysis shows how the three most notable models of consolation available today are in fact rooted in very different historical eras, highlighting correspondingly different concerns. The rejection of some aspects of consolation can thus go hand in hand with the fact that the psychological literature on grief and bereavement retains important aspects of consolation. With its strong emphasis on the *longue durée* of ideas about consolation, the Four-Axis Model can be useful for the interpretation of a wide range of spatial and historical phenomena, and Jedan offers examples of how the model can be applied, making use of Maddrell's tripartite model of consolation spaces.

Avril Maddrell's chapter focuses on bereavement and draws on previous scholarship in emotional-affective geographies of loss and remembrance to explore the dynamic and dialectical relationship between grief and consolation. It outlines a conceptual framework for understanding the *spatial* experience and practices associated with loss and remembrance, and how these in turn can be experienced – and even curated – as spaces and practices of consolation as part of wider therapeutic spatialities. This framework is used to 'map' and analyse the ways in which varied relational factors are expressed and experienced in and through particular places, including the body, home, landscapes and the virtual arenas of online memorials, social media and faith communities. This framework is then used to understand something of the bittersweet comfort which kin report experiencing as a result of deceased organ donation and the resulting consolation of knowing beneficiaries of donated organs are 'living memorials', embodying a degree of corporeal liveliness of the deceased, and in some cases engendering a sense of biologically grounded kinship.

The second section, which analyses spatial-conceptual constellations in Europe, the historical breeding ground of much of today's discussion about consolation, opens with two chapters based on case studies in the Netherlands (Nugteren, Hoondert). The other two chapters discuss case studies from England and Scotland (MacKinnon) and Denmark (Kjærsgaard). In 'Consolation and the "Poetics" of the Soil in "Natural Burial" Sites', Albertina Nugteren draws on her ethnographical monitoring of three natural burial sites in the south-east of the Netherlands. Nugteren distinguishes and discusses consolation in relation to three motivations or 'catalysts' of a rising interest in natural burial: (1) environmental concerns to minimise one's ecological footprint; (2) spiritual and aesthetic needs to 'get back to nature' in death; and, most radically, (3) a counter-cultural rejection of 'ingrained divides between body and soul, or between humans and nature'. Whilst the first two catalysts are compatible with 'shallow' ecological outlooks that still put the individual centre stage, the third derives from a sense of 'ontological alienation'. Natural burials motivated by such radical concerns focus on the soil. Nugteren argues that the difference in perspective can account for the disturbing quality which natural burials possess for some: Focusing on the soil counteracts the 'cultural preference for upwardly mobile symbols'. The consolation found in the most radical form of motivation for natural burial should be understood as a 'biophilic surrender'.

In his chapter 'The Crematorium as a Ritual and Musical Consolationscape', Martin Hoondert combines fieldwork at the Tilburg crematorium with music theory to focus on the under-theorised role of music in ritual studies. He argues that music creates space in interaction with the ritual and physical place and that the consolationscape produced by music must be understood in the light of the interaction of the three categories. For Hoondert, it is an important argument that the 'romantic' music – the label is here not understood as a historical term, but denotes a musical style ('slow tempo, smooth rhythms and relatively few dissonant harmonies') – often heard at cremations is not intrinsically 'more capable of bringing about a feeling of consolation than other styles'. If it functions as 'comfort' music *par excellence* that attempts to transform the listeners' mood, this must be 'strongly determined by the physical and ritual space'.

In 'Emotional Landscapes' Dolly MacKinnon analyses two memorial sites for victims of civil war in seventeenth-century England and Scotland. She approaches emotions from a framework of 'emotional practices' pioneered by Monique Scheer (2012). Emotional practices must be embedded in 'emotional communities' which actively mobilise, name, communicate and regulate emotions. Approaching memorial sites through a lens of emotional practices and communities allows MacKinnon to trace the history of two exemplary sites, Naseby (Northamptonshire) and Wigtown (Dumfries and Galloway), and to lay bare important aspects of memorialisation and consolation: first, how access to memorials may have been regulated by, and restricted to, closely knit emotional communities, as is witnessed by the fact that the early Naseby memorials were placed on private land, and access thus was restricted to like-minded groups of Royalists. Second, the degree to which such sites can be contested; the memory that is consoling to descendants of the victims can be a cause of triumph for those aligning themselves with the victorious. Even the existence of some groups of victims may be controversial, as happened in the curiously gendered denial of female victims of the Wigtown executions, and the denial of a 'baggage train' of women and children in the case of the Battle of Naseby.

In the chapter 'Danish Churchyards as Consolationscapes', Anne Kjærsgaard questions the highly secular appearance of Danish cemeteries. Indeed, an interpretation of Danish cemeteries as secular does not tally with the fact that most are run by the established Danish Lutheran church. Kjærsgaard argues that the visual restraint and absence of Christian symbols should not be understood as 'secular' but as the materialisation of Protestant aesthetic-religious norms. She argues further that the relatively small number of publicised contestations of those Protestant norms allow us to infer a much larger number of conflicts that have been quietly dealt with by local church authorities. The examples provided by Kjærsgaard identify Danish churchyards as consolationscapes that are the backdrop to the rich ritual activities of lived religion. On this basis, Kjærsgaard asks whether Denmark's Protestant churchyards are, or can continue to be, sufficiently inclusive.

The third section shifts the boundaries of discussion beyond the Global North. It opens with anthropological research in Australia, among the Tiwi on Melville Island and Bathurst Island, approximately 80 km north of

Darwin. In 'Moving through the Land', Eric Venbrux analyses the nexus between ritual, consolation and space. He focuses on rituals of wailing and lament to show how ritual drama can provide consolation. He thus counteracts the comparative neglect of consolation through ritual activity in the study of Aboriginal Australia, which has led to remarkable misinterpretations, denying the Tiwi mourners, for instance, their sincere feelings of grief. Venbrux reconstructs the Tiwi cycle of mortuary rites in which territorial passage, wailing and consolation are interlinked. Ideally, the rites start in the localities of the living and with intervals go on in space and time until the beginning of the final rite (called *iloti*, meaning 'for good') at the burial place, an area reserved for the spirits of the dead. In the ritual drama, given its purpose to direct the spirit of the dead from the world of the living to the world of the dead, the people of different bereavement status all play a role in the remembrance and dissolution of a particular metaphorical relationship with the deceased. In this context compassionate support and protection is given to the bereaved.

The other chapters in this section focus on Africa, more precisely what UN publications still refer to as 'Sub-Saharan Africa', i.e. Zimbabwe, Uganda and Senegal. The chapters by Joram Tarusarira and Sophie Seebach both put the issue of consolation in post-conflict societies centre stage. Both suggest to us editors that in the wake of continuing-bonds research in recent years, death studies researchers have *decathected* from Sigmund Freud in too radical a way, since both fruitfully employ parts of the analytical vocabulary of Sigmund Freud and his followers to characterise ritual and consolation.

In 'Rituals, Healing and Consolation in Post-conflict Environments', Tarusarira analyses the consolatory function of Ndebele funeral rites, with special attention to *umbuyiso*, a ritual 'bringing home' one year after the initial funeral. He attributes an important role to 'transitional objects' and to 'closure' produced by the rituals, and he analyses how a state-sponsored massacre known as Gukurahundi, which took place between 1982 and 1987 among the Ndebele people, and its legacy of disappeared persons hinders the performance of ritual and the finding of consolation.

In 'Love the Dead, Fear the Dead', Sophie Seebach focuses on death rituals in Acholi in northern Uganda, a region similarly torn by civil war. The 'LRA war', so called after Joseph Kony's 'Lord's Resistance Army' the most successful rebel army opposing central government, raged from 1986 to 2006, with deportations, abductions of children, killings and the mass encampment of civilians. Seebach shows not only how the civil war hindered the population's ability to perform fitting death-rites, but also how in the post-conflict era reburials integrate the dead into the traditional ritual texture and thus offer consolation. Seebach analyses how death rituals produce consolatory landscape, but she also points to the ambivalences of continuing bonds under-theorised in the recent literature. In this context she reminds us of the usefulness of Freud's notion of the 'uncanny': just as the compound needs to be safeguarded against the intrusion of the uncanny bush, just so 'people might (...) feel trapped in the continuing bonds between the living and the dead' and feel the urge to push back, so much so that one of Seebach's interviewees commented on Danish burial rites 'Ah, your custom [of cremation] is better'.

'It's God's Will', the final chapter in this collection, analyses consolation and religious meaning-making in urban Senegal. With the great majority of its population identifying as Muslim, Senegal adds an interesting perspective to the previous two African case studies. As highlighted by Ruth Evans, Sophie Bowlby, Jane Ribbens McCarthy, Joséphine Wouango and Fatou Kébé, geographies of loss in majority Muslim contexts are an understudied area. Their empirical work is particularly wideranging, drawing on fifty-nine in-depth interviews with individuals who had recently experienced an adult relative's death, working moreover with key informants, focus groups and participatory workshops. The authors interpret their findings on the basis of Dennis Klass's (2014) framework for religious consolation, and underscore *inter alia* the importance of continuing bonds for consolation, and of ritual activities as expressions of continuing bonds.

References

Allard, O. 2013. To cry one's distress: death, emotions and ethics among the Waro of the Orinoco Delta. *Journal of the Royal Anthropological Institute* 19/3, 545–561.

Alonso del Real, C. 2001. *Consolatio*. Navarra: Ediciones Universidad de Navarra.

Baltussen, H., ed. 2013. *Greek and Roman Consolations: Eight Studies of a Tradition and Its Afterlife*. Swansea: The Classical Press of Wales.

Behar, R. 1991. Death and memory: from Santa María del Monte to Miami Beach. *Cultural Anthropology* 6/3, 346–384.

Bondi, L., Davidson, J. and Smith, M. 2005. Introduction: geography's 'emotional turn', in *Emotional Geographies*, edited by J. Davidson, L. Bondi and M. Smith. Aldershot: Ashgate, 1–16.

Bregman, L. 2012. *Preaching Death: The Transformation of Christian Funeral Services*. Waco: Baylor University Press.

Buch Segal, L. 2016. *No Place for Grief: Martyrs, Prisoners, and Mourning in Contemporary Palestine*. Philadelphia: University of Pennsylvania Press.

Danforth, L. M. 1982. *The Death Rituals of Rural Greece*. Princeton: Princeton University Press.

Davies, D. J. 2002. *Death, Ritual and Belief: The Rhetoric of Funerary Rites*. Second edition. London and New York: Continuum.

Dunn, C. E., Le Mare, A. and Makungu, C. 2016. Connecting global health interventions and lived experiences: suspending 'normality' at funerals in rural Tanzania, *Social & Cultural Geography* 17/2, 262–281.

Favez, C. 1937. *La consolation latine chrétienne*. Paris: Vrin.

Gamliel, T. 2014. *Aesthetics of Sorrow: The Wailing Culture of Yemenite Jewish Women*. Detroit: Wayne State University Press.

Geertz, C. 1973. Religion as a cultural system, in *The Interpretation of Cultures*. New York: Basic Books, 87–125.

Geertz, C. 1986. Making experience, authoring selves, in *The Anthropology of Experience*, edited by V. W. Turner and E. B. Bruner. Urbana: University of Illinois Press, 373–380.

Gilbert, S. M. 2006. *Death's Door: Modern Dying and the Ways We Grieve*. New York: Norton.

Gillis, C. M. 2012. Seeing differently: place, art and consolation, in *Emotion, Identity and Death: Mortality across disciplines*, edited by D. J. Davies and C.-W. Park. Farnham: Ashgate, 99–108.

Glare, P. G. W., ed. 1996. *Oxford Latin Dictionary*. Reprinted with corrections. Oxford: Oxford University Press.

Goss, R. E. and Klass, D. 2005. *Dead but Not Lost: Grief Narratives in Religious Traditions*. Walnut Creek: AltaMira Press.

Gregg, R. C. 1975. *Consolation Philosophy. Greek and Christian Paideia in Basil and the Two Gregories*. Cambridge, MA: The Philadelphia Patristic Foundation.

Hockey, J., Komaromy, C. and Woodthorpe, K., eds. 2010. *The Matter of Death: Space, Place and Materiality*. Basingstoke: Palgrave.

Holloway, P. A. 2001. *Consolation in Philippians: Philosophical Sources and Rhetorical Strategy*. Cambridge: Cambridge University Press.

Jedan, C. 2014a. Troost door argumenten: Herwaardering van een filosofische en christelijke traditie, in *Dood en troost*, edited by C. Jedan and E. Venbrux. Zoetermeer: Boekencentrum [= *Nederlands Theologisch Tijdschrift* 68/1&2], 7–22.

Jedan, C. 2014b. De Grieks-Romeinse consolatio, in *Dood en troost*, edited by C. Jedan and E. Venbrux. Zoetermeer: Boekencentrum [= *Nederlands Theologisch Tijdschrift* 68/1&2], 165–173.

Jedan, C. 2017. The rapprochement of religion and philosophy in ancient consolation: Seneca, Paul, and beyond, in *Religio-Philosophical Discourses in the Mediterranean World*, edited by G. H. van Kooten and A. K. Petersen. Leiden/Boston: Brill, 159–184.

Jedan, C. and Venbrux, E., eds. 2014. *Dood en troost*, Zoetermeer: Boekencentrum [= *Nederlands Theologisch Tijdschrift* 68/1&2].

Johann, H.-T. 1968. *Trauer und Trost. Eine quellen- und strukturanalytische Untersuchung der philosophischen Trostschriften über den Tod*. Munich: Fink.

Kassel, R. 1958. *Untersuchungen zur griechischen und römischen Konsolationsliteratur*. Munich: Beck.

Kellaher, L. and Worpole, K. 2010. Bringing the dead back home: urban public spaces as sites for new patterns of mourning and memorialisation, in *Deathscapes: New spaces for death, dying and bereavement*, edited by A. Maddrell and J. Sidaway. Farnham: Ashgate, 161–180.

Kjærsgaard, A. 2017. *Funerary Culture and the Limits of Secularization in Denmark*. Zürich: LIT.

Klass, D. 2013. Sorrow and solace: neglected areas in bereavement research. *Death Studies* 37/7, 597–616.

Klass, D. 2014. Grief, consolation, and religions: a conceptual framework, *Omega: Journal of Death and Dying* 69/1, 1–18.

Klass, D. (unpublished). The nature of religious consolation for the bereaved. Unpublished draft. Available at: http://www.academia.edu/2247617/The_Nature_of_Religious_Consolation_for_the_Bereaved. Accessed 11 July 2018.

Klass, D., Silverman, P. R. and Nickman, S. L., eds. 1996. *Continuing Bonds: New Understandings of Grief*. London: Routledge.

Klass, D. and Steffen, E. M., eds. 2017. *Continuing Bonds in Bereavement: New Directions for Research and Practice.* London and New York: Routledge.

Kolcaba, K. 2003. *Comfort Theory and Practice: A Vision for Holistic Health Care and Research.* New York: Springer.

Kong, L. 1999. Cemeteries and columbaria, memorials and mausoleums: narrative and interpretation in the study of deathscapes in geography. *Australian Geographical Studies* 37, 1–10.

Kwon, H. 2013. *After Massacre: Commemoration and Consolation in Ha My and My Lai.* Berkeley: University of California Press.

Langenhorst, G. 2000. *Trösten lernen? Profil, Geschichte und Praxis von Trost als diakonischer Lehr- und Lernprozeß.* Ostfildern: Schwabenverlag.

Laqueur, T. W. 2015. *The Work of the Dead: A Cultural History of Mortal Remains.* Princeton: Princeton University Press.

Lillo Redonet, F. 2001. *Palabras contra el dolor: La consolación filosófica Latina de Cicerón a Frontón.* Madrid: Ediciones Clásicas.

Lipset, D. and Silverman, E. K., eds. 2016. *Mortuary Dialogues: Death Ritual and the Reproduction of Moral Community in Pacific Modernities.* New York: Berghahn.

Lutz, C. A. 1998. *Unnatural Emotions: Everyday Sentiments on a Micronesian Atoll and Their Challenge to Western Theory.* Chicago: University of Chicago Press.

Maddrell, A. 2009a. Mapping changing shades of grief and consolation in the historic landscape of St. Patrick's Isle, Isle of Man, in *Emotion, Culture and Place*, edited by M. Smith, J. Davidson, L. Cameron and L. Bondi. Ashgate: Aldershot, 35–55.

Maddrell, A. 2009b. A place for grief and belief: the Witness Cairn at the Isle of Whithorn, Galloway, Scotland. *Social and Cultural Geography* 10, 675–693.

Maddrell, A. 2011. Bereavement, belief and sense-making in the contemporary British landscape: three case studies, in *Emerging Geographies of Belief*, edited by C. Brace, D. Bailey, S. Carter, D. Harvey and N. Thomas. Cambridge: Cambridge Scholars, 216–238.

Maddrell, A. 2012. Online memorials: the virtual as the new vernacular. *Bereavement Care* 31, 46–54.

Maddrell, A. 2013. Living with the deceased: absence, presence and absence-presence. *Cultural Geographies* 20/4, 501–522.

Maddrell, A. 2016. Mapping grief: a conceptual framework for understanding the spatialities of bereavement, mourning and remembrance. *Social and Cultural Geography* 17/2, 166–188.

Maddrell, A. and Sidaway, J. 2010. Introduction: Bringing a spatial lens to death, dying, mourning and remembrance, in *Deathscapes: New spaces for death, dying and bereavement*, edited by A. Maddrell and J. Sidaway. Farnham: Ashgate, 1–16.

Manning, C. E. 1974. The consolatory tradition and Seneca's attitude to the emotions. *Greece & Rome* 21, 71–81.

McClure, G. W. 1991. *Sorrow and Consolation in Italian Humanism.* Princeton: Princeton University Press.

Moos, P. von. 1971–1972. *Consolatio. Studien zur mittellateinischen Trostliteratur über den Tod und zum Problem der christlichen Trauer.* 4 vols. Munich: Fink.

Mueggler, E. 2017. *Songs for Dead Parents: Corpse, Text and World in Southwest China.* Chicago: Chicago University Press.

Norberg, A., Bergsten, M. and Lundman, B. 2001. A model of consolation. *Nursing Ethics* 8/6, 544–553.

Price, L., McNally, D. and Crang, P. eds. 2018. *The Geographies of Comfort.* London and New York: Routledge.

Resch, C. 2006. *Trost im Angesicht des Todes. Frühe reformatorische Anleitungen zur Seelsorge an Kranken und Sterbenden.* Tübingen and Basel: Francke.

Rittgers, R. K. 2012. *The Reformation of Suffering: Pastoral Theology and Lay Piety in Late Medieval and Early Modern Germany.* Oxford: Oxford University Press.

Robinson, J. D. 2010. Lethean landscapes: forgetting in late modern commemorative spaces, in *Memory, Mourning, Landscape*, edited by E. Anderson, A. Maddrell, K. McLoughlin and A. Vincent. Amsterdam and New York: Rodopi, 79–97.

Rosaldo, R. 2014. *The Day of Shelley's Death: The Poetry and Ethnography of Grief.* Durham: Duke University Press.

Roth, U. 2002. *Die Beerdigungsansprache: Argumente gegen den Tod im Kontext der modernen Gesellschaft.* Gütersloh: Kaiser.

Rugg, J. 2018. Consolation, individuation and consumption: towards a theory of cyclicality in English funerary practice, *Cultural and Social History* 15/1, 61–78.

Scheer, M. 2012. Are emotions a kind of practice (and is this what makes them have a history)? A Bourdieuian approach to understanding emotion. *History and Theory* 51, 193–220.

Scheper-Hughes, N. 1992. *Dead Without Weeping: The Violence of Everyday Life in Brazil.* Berkeley: University of California Press.

Scourfield, J. H. D. 1993. *Consoling Heliodorus: A Commentary on Jerome, Letter 60.* Oxford: Oxford University Press.

Seremetakis, N. C. 1991. *The Last Word: Women, Death, and Divination in Inner Mani.* Chicago: University of Chicago Press.

Simonds, W. and Rothman, B. K. 1992. *Centuries of Solace: Expressions of Maternal Grief in Popular Literature.* Philadelphia, PA: Temple University Press.

Solomon, R. C. 1984. Getting angry: the Jamesian theory of emotion in anthropology, in *Culture Theory: Essays on Mind, Self, and Emotion*, edited by R. A. Shweder and R. A. LeVine. Cambridge: Cambridge University Press, 238–254.

Stevenson, O., Kenten, C. and Maddrell, A. 2016. Editorial: And now the end is near: enlivening and politicising the geographies of dying, death and mourning. *Social and Cultural Geography* 17/2, 153–165.

Sykes, J. B., ed. 1982, *The Concise Oxford Dictionary.* Oxford: Clarendon Press.

Tarlow, S. 1999. *Bereavement and Commemoration: An Archaeology of Mortality.* Oxford: Blackwell.

Venbrux, E. 1991. A death-marriage in a Swiss mountain village. *Ethnologia Europaea* 21/2, 193–205.

Weichselgartner, J. and Kelman, I. 2015. Geographies of resilience: challenges and opportunities of a descriptive concept. *Progress in Human Geography* 39, 249–267.

Weyhofen, H.-T. 1983. *Trost: Modelle des religiösen und philosophischen Trostes und ihre Beurteilung durch die Religionskritik.* Frankfurt: Peter Lang.

Woods, R. 2006. *Children Remembered: Responses to Untimely Death in the Past.* Liverpool: Liverpool University Press.

Part I
Reviving consolation

1 What is consolation?

Towards a new conceptual framework

Christoph Jedan

Introduction

'Consolation is grief's traditional amelioration, but contemporary bereavement theory lacks a conceptual framework to include it' (Klass unpublished). I would go even further and state that, over the past two centuries, Western culture as a whole has become increasingly suspicious of the concept of consolation. To many people, the word 'consolation' seems to suggest a questionable religious cultural baggage and a want of valued attributes such as activity and resilience. Little wonder that other concepts, such as 'coping', are now used far more often to prescribe ways of dealing with loss.[1]

Yet we can ill afford to abandon the concept of consolation and closely related notions such as 'comfort' and 'solace'. They denote experiences, attitudes and activities that no other words can adequately identify. We cannot, for instance, 'find coping' during a long forest walk. If, as Kant argued, perceptions are 'blind without concepts', we need the concept of consolation to adequately experience and analyse *consolationscapes* – the spatially situated phenomena and practices to do with loss and grief.

Owing to neglect of the concept of consolation, it is difficult to find convincing analyses of what consolation actually is. The extant conceptual frameworks are not only few and far between but they are derived from limited cultural-historical sources and thus offer narrow views of the richness of the concept. This is a danger inherent in death studies more generally: once longer-term trends disappear out of sight, recent phenomena acquire the reputation of being unprecedented and markers of a new era (whether late-modern, neo-modern, postmodern, or contemporary). What is needed, however, is the corrective provided by a 'long view'. Without downplaying genuine cultural change, we must acknowledge that in respect of consolation there are both undeniable continuities and also reversals to older forms. For instance, how does the recent emphasis on biographies of the deceased (Walter 1996) relate to the emphasis on biography that we find since the late eighteenth century, and to

---

1 As the comparison of n-grams of the concepts of consolation and coping can show (Jedan 2017b, utilising the Google 'Ngram Viewer': https://books.google.com/ngrams.

even earlier emphases on biography in early modern eulogies or ancient Greco-Roman consolations? Is the emphasis on earthly life a recent phenomenon (Davies 2005, 2008), or has it been part and parcel of earlier consolatory efforts, too? Taking the long view is paramount for the analysis of the spatialities of consolation, not least because spatial arrangements themselves are often quite old: a war memorial or a cemetery might date from the nineteenth century, and we might want to know how the experience of consolation today relates to the consolatory purposes of their creators.

In the present chapter, I introduce the Four-Axis Model, a conceptual framework of consolation that aspires to that much-needed 'long view' and thus, I hope, to a higher degree of generality than previous attempts. It will be evident from the following pages that my own conceptual work has taken its cue from written sources, particularly from dedicated consolatory texts and treatises that, for better or worse, focus on death as the most significant form of loss (Jedan 2014b, 2017a, 2017b). It seems to me, however, that the framework can also be applied to other types of loss. All that is required is to replace 'death' with 'loss' in the definition of consolation offered below. In addition, I hope it will be evident that the present chapter attempts a careful triangulation with other sources and materials to offer a conceptual framework that is genuinely useful for the interpretation of specific phenomena.

The chapter first presents three notable models of consolation. The frameworks are limited in that they represent different aspects of consolation. Their value lies in pointing (perhaps unwittingly) towards three different 'strands' or types of consolation: 'metaphysical and moral consolation', '(auto)biographical memorialisation', and 'professionalised consolation'. These strands, I argue, can be distinguished in the historical material, and their origins can be traced to different historical eras. Identifying recurrent themes, the Four-Axis Model is then developed against the backdrop of the historical variation. Finally the chapter discusses the application of the framework to selected spatial phenomena and possible limitations of the framework.

Before we begin in earnest, a word on terminology is in order. The availability in the English language of two closely related concepts, 'grief' and 'mourning', has led numerous scholars down the garden path of trying to differentiate between the two, for instance by claiming that 'grief' is about something internal to the subject (affective), whereas 'mourning' designates the social expression and ritualisation of inner grief. Klass (2014) has noted the questionable nature of differentiation between allegedly pre-social affects and their social cultivation and expression; at any rate, other languages such as German ('Trauer') or Dutch ('rouw') offer no linguistic support for it. In this chapter, I use 'grief' and 'mourning' interchangeably.

Three notable models of consolation

The first model of consolation in this review was formulated by Weyhofen (1983: 249):

Consolation's point of departure is a difference – more specifically, a contradiction – between the human and the world. On one side we find the interests, wishes and goals of the human being; on the other side there is the world, which does not comply with those interests, wishes and goals. Consolation is the answer to the suffering that is caused by that difference. The goal of consolation is to remove the difference and to produce a reconciling identity.[2]

According to Weyhofen, the difference can be removed in two ways: first, by altering one's interests, wishes and goals – whether by aligning them with 'the world' as it is (the Stoic solution) or, more radically, by a mystical renunciation of all interests, wishes and goals – and, second, by hope for an eschatological transformation of the world when later, perhaps in an afterlife, the world will finally be congruent with one's wishes (Weyhofen 1983: 249–252). The definition has a clear emphasis on metaphysics and religion: consolation is to address a metaphysical and religious suffering.

The second model of consolation to be discussed stems from a team at Umeå University's Department of Nursing led by Astrid Norberg. Focusing on the concepts of 'alienation' and 'communion', it leans heavily on Gabriel Marcel's work. Marcel postulates 'that deep in human beings there is a universal brotherhood that is revealed through a global feeling of a "we" (communion)' (Norberg et al. 2001: 550). In this model, suffering is an existential loneliness, an alienation of the suffering persons 'from themselves, from other people, from the world, and from their transcendent source of meaning' (2001: 544). Consolation involves the reconstitution of communion, 'a changed perception of the world in suffering persons' (2001: 544) in which the suffering Other is 'acknowledged as a presence' (2001: 551). This can happen in silence, but an important role is attributed to a dialogue in which the suffering person can express their suffering and share it with the other person.

The third model was proposed by Klass (2014) to address the apparent inability of contemporary bereavement theory to embrace the concept of consolation. The reason for this inability is the mistaken method of 'psychological individualism' (2014: 7), which ignores the inter-subjective social aspect of grief and consolation. In stressing the aspect of consolation as communion, Klass takes his cue from the use of existentialist philosophy by Norberg, Bergsten and Lundman (2001), as well as Walter's (1996) insistence on the importance of 'conversations between those who knew the deceased' for contemporary practices of bereavement (2014: 5):

> Grief is a social emotion and thus is inter-subjective, even at the level of biological response. Grief is an interaction between interior, interpersonal, communal, and cultural narratives that are charged with establishing the meaning of the deceased's life and death. Consolation happens

---

2   Here and elsewhere, unless stated explicitly, the translations are my own.

in the same inter-subjective space as grief. Consolation soothes and alleviates the burden of grief, but does not take away the pain. Consolation is trust in a reality outside the self.

Not fully integrated with his model of consolation but nonetheless important is Klass's (2014: 2–3, 13) two-fold suggestion (1) that religions as important loci of consolation offer (a) an encounter or merger with transcendent reality, (b) a worldview in which our individual life narratives are nested and, finally, (c) a community in which the transcendent reality, our worldview, and our own experience are validated; and (2) that (non-religious) self-help groups recreate the aforementioned characteristics of religion to sustain and validate continuing bonds with the deceased that nevertheless lack acceptance in contemporary grief theory and culture.[3]

Clearly, the three models of consolation stress different points: Weyhofen emphasises a metaphysical or religious perspective; Norberg, Bergsten and Lundman highlight the need to overcome a sense of isolation on the part of the bereaved; Klass's most characteristic addition to Norberg, Bergsten and Lundman is his insistence on the shortcomings of contemporary grief theory. I want to suggest that the three models with their characteristic emphases point to three different types or strands of consolation that rose to prominence in the specific cultural contexts of different eras. Whilst Weyhofen's idea of consolation ties in with the *metaphysical and moral consolation* devised in the premodern era, Norberg, Bergsten and Lundman's concept of consolation is predicated on the increasing importance of the individual and the concomitant sense of existential isolation in the modern era that was answered with the consolatory strategy of *(auto)biographical memorialisation*. Finally, Klass reactivates premodern and modern consolatory strategies in his opposition to a *professionalised consolation* that rose to prominence after WWI.

Three strands of consolation

In this section I shall describe the three important 'strands' or 'types' of consolation already mentioned: 'metaphysical and moral consolation', '(auto)biographical memorialisation' and 'professionalised consolation'. These strands have gained prominence in very different cultural and historical circumstances.

3   For lack of space, I have restricted my discussion to the above three models. Other concepts that have direct bearing on the present inquiry include earlier publications by Dennis Klass on the role of religion in bereavement (e.g. Klass 1993; 2006; 2013), constructivist models of grief (e.g. 'assumptive world', 'relearning the world', 'meaning reconstruction': Parkes and Prigerson 2010; Attig 2011; Neimeyer, 2001; Kauffman 2002); the suggestion that ritual, literary, musical responses to death can be characterised as 'words against death' (Davies 2002, see Jedan 2014a), and analyses of the rhetorical structures of eulogies (Kunkel and Dennis 2003).

In this sense they follow historically upon one another, but their use extends to the present day so that they now co-exist.

Inevitably, such 'strands' or 'types' are abstractions and simplifications. In this respect, they function in much the same way as sociology's 'ideal types' (e.g. Walter 1994), and they are threatened by similar objections to those levelled against histories of mentalities (e.g. Ariès 1981). I have taken care to pre-empt such critiques as far as possible, first by identifying as much as I can the historical sources in which the new developments become visible, and second by explicitly recognising that consolation in itself is a regulatory and thus also adversarial or agonistic phenomenon: consolation is part of a cultural attempt to regulate or 'police' grief (Walter 1999). The interesting question is then: what is 'the Other' or who are 'the Others' of consolation? Consolation combats an excess of grief that is presented as unhelpful, dangerous and destructive, and presents as its opposite an *ideal of acceptable grief*. Little wonder, then, that the types of consolation are themselves vulnerable to *specific cultural threats*. Whilst the metaphysical and moral consolation that can be traced back to premodern cultures was (and still is) threatened by attacks on its metaphysical and religious underpinnings, and perhaps even more so by the perceived insufficiency of general prescription, modern (auto)biographical memorialisation has come under attack from detractors of individuality in the name of ideas as diverse as the nation or the scientific world-view. Lastly, the professionalised consolation of the past hundred years has been challenged primarily by reference to personal experience, repeating motifs established in the (auto)biographical, memorialising consolation of the modern era as well as in the metaphysical and moral consolation of the premodern era.

On the basis of a certain abstraction, however, we can also identify *typical genres* in which the strands of consolation are expressed. In addition to important early texts representing a consolatory strand, we can identify *iconic texts* that remain important points of reference. Finally, the strands of consolation also contain *depictions of a fulfilled or intensified life*, which function essentially as interpretative templates for rewriting the biography of the deceased to show that his or her life was worthwhile and need not cause (vicarious) regrets.

Table 1.1 provides an overview of the three strands. A more detailed analysis of the strands and the sources underpinning my interpretations is given below.

1 Premodern era: Metaphysical and moral consolation

We know little about the exact beginnings of consolatory writing in the Ancient Greco-Roman culture. It is very probable that such reflections grew out of the rhetorical practice of delivering eulogies. Antiphon, an orator of the fifth century BCE, is reported to have invented and marketed an 'anti-grief technique' (*technê alupias*),[4] and in the fourth century BCE Plato wrote

---

4   'But while he was still busy with poetry he invented a method of curing distress, just as physicians have a treatment for those who are ill; and at Corinth, fitting up a

Table 1.1 Three strands of consolation

	Metaphysical and moral	Biographical memorialisation	Professionalised therapy
Era of invention	Premodern	Modern	Twentieth century/ contemporary
Typical genre	Treatise, sermon	(Auto)biography, memoir	Psychological handbook
Important early text	Crantor, *On Grief* (c. 300 BCE)	[Adams], *Agnes and the Little Key* (1837)	Freud, *Mourning and Melancholia* (1917)
Iconic text	Boethius, *Consolation of Philosophy* (524 CE)	C. S. Lewis, *A Grief Observed* (1961)	Worden, *Grief Counselling and Grief Therapy* (1983–2010)
Ideal of fulfilled life	Virtue	Uniqueness of individual	Completion of discrete tasks; bucket list
Ideal grief	Minimised/ suppressed	Aestheticised; residual melancholy	Publicly invisible/ disclosed to selected confidant(e)s
Opposed cultural trends	Scepticism about metaphysical/religious framework; insufficiency of general doctrines for individual case	Precedence of community over individual (nation, state, large-scale loss of life); scientific world-view challenges individual	Personal experiences and world-views challenge neutrality of scientific world-view

dialogues such as the *Apology* and the *Phaedo* that touched upon questions of death and grief. However, the first focused and sustained reflection on consolation that we know of was written almost a century later, when the Platonist philosopher Crantor (*c.* 335–275 BCE) wrote a treatise *On Grief* (*Peri penthous*). *On Grief* proved enormously influential. It was used in the construction of later consolations and was referred to throughout Antiquity. The treatise was subsequently lost, but there are enough traces of it in the later consolatory literature for classicists to have attempted reconstructions of Crantor's work.[5] Crantor appears to have argued *inter alia* that mourning is useless, since it cannot undo the loss that has occurred, and moreover that it is harmful for the bereaved. Crantor thus views grief in negative terms: it is dangerous, destabilising, and unwanted. Grief is an emotion one wants to get rid of as much as possible, even if different ancient consolers disagreed over

room near the market-place, he wrote on the door that he could cure by words those who were in distress; and by asking questions and finding out the causes of their condition he consoled those in trouble. But thinking this art was unworthy of him he turned to oratory' ([Plutarch], *Lives of the Ten Orators* 833C [Fowler 1936: 351]).

5　See e.g. Kassel (1958), Johann (1968) and Mette (1984).

the extent to which this was possible. Crantor and his followers underpinned their exhortations by a metaphysical and moral view of nature in which death has a legitimate place: the world is presented as a stage on which human lives unfold in a play about the perfection of character. The ancient consolers thus advocated a *virtue-focused* ideal of the fulfilled life, a *moral* intensification of life. Mortality is inscribed in every role of the play: one needs to accept that as a fact. The goods of life are independent of its duration; ultimately, death must be considered a release from the evils of the physical world. Whilst such consolations regularly sketched attractive afterlives, those ideas were not an essential element: one of the inspirational texts for ancient consolation, Plato's *Apology of Socrates*, had already offered the alternative of death as either a journey to a heavenly place or a complete annihilation, and stipulated further that neither scenario was a ground for fear. Subsequent ancient consolers, including the often-lambasted Epicurus, remained faithful to this Platonic line (Jedan 2014a, 2017a). And whilst many but not all of the ancient consolations were religious in outlook, not even the religious ones uniformly attributed a role to God (or the gods) in the afterlife. A 'merger with transcendent reality' (Klass 2014) is, in short, not constitutive for metaphysical and moral consolation.

To us today, the Ancient Greco-Roman consolatory advice might seem harsh, but it allows us to assess the strength of its cultural adversary. In premodern culture there seems to have been a strong tendency towards exuberant grief. Many of the ancient consolers 'other' this exuberant grief as vulgar and even point to legal restrictions intended to curb excessive displays of grief. Clearly, the 'correct' attitude towards grief was also advocated in terms of social differentiation.[6]

The preferred genre of this type of consolation includes treatises and sermons, outlining a therapeutic world-view that helps the bereaved to understand and accept death as a legitimate event against which they should not fight.

The rise of Christianity had less of an impact on the nature of consolation than might be supposed. There are far-reaching similarities between 'philosophical' and 'Christian consolers', which make it appropriate to speak of 'metaphysical and moral consolation', thus embracing both philosophical and theological consolatory modes. Good examples are Seneca's *To Marcia on Consolation* and Paul of Tarsus' *Letter to the Philippians*, each written around the middle of the first century CE. Ostensibly letters, both are essentially treatises or sermons outlining a worldview from which the bereaved can draw support. There are remarkable intellectual commonalities between the two texts, which have led commentators to speculate that Paul might have known

6   See e.g. Plutarch, *Consolation to His Wife* 609E–610B (De Lacy and Einarson 1959: 591–593). It is worth pointing out that negative views of grief, and injunctions to moderate grief continue to this day, as is underscored by strong reactions against the 'policing' of grief and against expectations of 'closure' (see below). At least in the English culture, Adam Smith (1759/2000) may have played a pivotal role for continuing injunctions of grief moderation.

and sought to emulate Greco-Roman philosophical consolation (Holloway 2001). Both Seneca and Paul opt for a tight regulation and even eradication – a less favourable interpretation would be suppression – of grief: while Paul exhorts his addressees that 'you should not grieve like the rest of mankind, who have no hope' (1 Thess. 4:13 REB), Seneca charges his addressee to 'Blush to have a low or common thought, and to weep for those dear ones who have changed for the better!' (*To Marcia on Consolation* 25.3 [Basore 1932: 91]). Both recommend a life of virtue in this world as an antidote to the destructive power of death, Seneca because the possession of virtue makes life complete and will secure for the virtuous an exalted afterlife, and Paul because virtue is demanded and will be rewarded by his heavenly father.

If we wished to identify a single iconic text that ideally represents this type of consolation it would have to be Boethius' *Consolation of Philosophy* (524 CE; ed./trans.: Tester 1973), the most-frequently reprinted of all the ancient consolations. Boethius, himself a Christian, consoles himself for his fall from the emperor's grace and impending sentence of death. The consonance he appears to see between philosophy and theology makes it unnecessary for him to highlight Christian motifs: Neoplatonic metaphysics yields the assurances he requires, and his consolation is administered not by Christ, but by a personified Philosophy in a mixture of dialogues and verses.

Whilst the Boethian consolation is characterised by its 'ecumenical' metaphysical character embracing philosophy and theology, the Middle Ages witnessed an increasing emphasis on specifically Christian motifs of consolation: gradually, consolation becomes 'Christianised', but the consolers continue to take their cue from Boethius and Neoplatonism. This may be seen in, for instance, Meister Eckhart's *Book of Divine Consolation* (c. 1308–1318, ed./ trans.: Quint 1979) and in particular Jean Gerson's pointedly entitled *Consolation of Theology* (c. 1418; trans.: Miller 1998, analysis: Burrows 1991). Gerson models his consolation on Boethius' alternation of prose and poetry, and expresses his intention to continue where his illustrious predecessor left off. The Christianisation of consolation is also visible in the emergence of a closely related form of literature, since Gerson's work *De arte moriendi* (part of his *Opus tripartitum, c.* 1414/c. 1477) appears to have stimulated the authoring of *Ars Moriendi* treatises, in effect ritual and spiritual self-help manuals on how to face death. Particularly well-known are the *Tractatus artis bene moriendi* (c. 1420) and a shorter version of it with woodcut depictions of deathbed scenes, juxtaposing the temptations of the devil and the benign inspirations of the angel of faith.[7] The Protestant Reformation may have further increased the trend towards a Christianisation of consolation, as can be seen in the *Ars Moriendi* treatises of Luther (1519, 1520) and in Erasmus of Rotterdam's response (1539), but it is certainly the continuation of a long-term trend.

---

7   Commonly referred to after their *incipits* as the CP and QS versions respectively. For sources and development of the *Ars Moriendi* literature, see O'Connor (1966).

To return to Gerson: his reputation was such that even the anonymously published, most-often reprinted and most widely read spiritual Christian book after the Bible was wrongly attributed to him: this was Thomas à Kempis' *Imitation of Christ* (*c.*1418; trans.: Wijdeveld 2003). The book contains a treatise in dialogue form that advocated an 'inner consolation' and a spiritual withdrawal from the world. With almost 4,000 reprints in 90 languages, the book's influence in fixing the public image of consolation as a Christian preoccupation and a matter of resignation can hardly be overrated.

If my interpretation is correct, it would be wrong to assume that cultural movements such as the Renaissance and Enlightenment simply and immediately shifted views of death and ushered in a decline of consolation (Choron 1963: 91–102; Stammkötter 1998). Decisive was the combination of a steady Christianisation of consolation and a gradual move away from attitudes of meekness, resignation and submission that became associated with consolation. The outcome was that the label 'consolation' became more and more toxic, even for efforts that clearly were consolatory in nature.[8]

All this, however, was a long-term development. In the premodern era, the biggest challenge to metaphysical and moral consolation was probably not anti-religious sentiment but an uneasiness as to whether highly abstract world-views sufficiently engaged individuals in mourning. Such uneasiness is already palpable in the ancient Greco-Roman consolations, where we find consolers going to great lengths to indicate that their advice was geared towards one specific loss. With the rising emphasis on the individual, new forms of consolation became inevitable.

Before we move on, however, it is worth emphasising that metaphysical and moral consolation has by no means disappeared. When contemporary scholars (e.g. Klass 2014: 6) emphasise the importance of a shared world-view and of narratives that 'place the life of the person who died and the lives of those who mourn into the context of cultural narratives about the meaning of the human condition at large', they essentially reiterate this 'premodern' strand of consolation.

2 Modern era: (Auto)biographical memorialisation

The most significant cultural change inaugurating what today we call 'modernity' was the growing sense of the importance of the individual. To appreciate the new role of the individual, one needs only to look at Goethe's 'Sturm und Drang' novel *The Sorrows of Young Werther* (1774/1985) which influenced the pan-European movement of Romanticism. This epistolary

8 In this cultural climate, the pointed recommendation of Epicurean or Stoic consolatory motifs, which belonged the extended family of metaphysical and moral consolation, would be understood as an attack on Christianity. D'Holbach, for instance, and Ludwig Feuerbach (1830/1960: 84–90) reverted to Epicurean consolatory motifs, Nietzsche (1889/1988: 113) lambasted the *Imitation* as 'effeminate' and recommended a largely Stoic attitude to loss and death.

novel was successful at an international level, and there was a veritable 'Werther fever' with a rise in suicides and imitations of Werther's dress code. Werther represents a new contrast between the individual and society, together with a sense of existential loneliness. The novel describes Werther's path from unfulfilled love to suicide, and suggests that the church's official ban on suicide was incompatible with the idea of God as a loving father (1774/1985: 107).

The growing importance attributed to the individual made the search for new forms of communion more necessary – and at the same time more difficult – than ever. Religion did not help in this regard, since there was a growing suspicion that religion might be no more than a cultural artefact.[9] Increased intimacy is one of the ways out of the impasse, and it is a strategy advocated in *The Sorrows of Young Werther* in spite of the disastrous consequences for its protagonist.

In this vein, the expression of emotions such as grief that the pre-modern metaphysical and moral consolation tended to view with suspicion were re-evaluated and even *cultivated*. Werther talks of 'sweet melancholy' and fondly remembers the 'tearful eye' of his beloved. Another important signal of the increased status of the emotions is the fact that religion could be described as based primarily on emotion (Schleiermacher 1799/1985). Immersion in nature is another strategy to find a communion for the individual, and it is also presented in *The Sorrows of Young Werther*. To make this possible, the picture of nature had to undergo a marked change. Whereas nature had been little more than a stage for human beings in the premodern metaphysical and moral consolation, nature was now emotionally charged; it was viewed with sentimentality as a mirror of the human personality.

The most important consolatory innovation of the era, however, was biography. Biography and memoirs must be regarded as the outstanding consolatory genres of the modern era. Of course, they were not entirely new genres. To focus on biographies, there had been sketches of lives in antiquity: Plutarch's *Parallel Lives* of famous persons had been literary models for centuries, and Greco-Roman metaphysical and moral consolations contained sketches of exemplars of coping (or not coping) with loss.[10] More recently Protestant funeral eulogies have customarily remembered the lives of the deceased (Sörries 2011: 107). However, such biographical sketches were primarily supplying exemplars for an ethical or religious stance; they were not focused on finding out what had made an individual human being 'tick'. The latter is the hallmark of biography in the modern sense.

Boswell's *Life of Johnson* (1791/2008), is often hailed as the first modern biography, and although it was not intended as a consolation, it offers us

---

9   For instance, Hegel (1796–97/1971: 236) and his circle of friends suggested that a properly Romantic religion should be 'the last and greatest work of mankind'.

10   Examples: *To Marcia on Consolation* 2–3 (Basore 1932: 9–17); Cicero, *Tusculan Disputations* 3.29–30 (King 1927: 261–63); [Plutarch], *A Letter of Condolence to Apollonius* 118D–119E (Babbitt 1928: 193–201).

valuable insights into the consolatory strand of (auto)biographical memorialisation. Boswell stresses the amount of material communicated by other people who knew Johnson, and he underscores the meticulous work that went into ascertaining the veracity of many details:

> I must be allowed to suggest, that the nature of the work, in other respects, as it consists of innumerable detached particulars, all which, even the most minute, I have spared no pains to ascertain with a scrupulous authenticity, has occasioned a degree of trouble far beyond that of any other species of composition. (...) Let me only observe, as a specimen of my trouble, that I have sometimes been obliged to run half over London in order to fix a date correctly.
>
> (1791/2008: 4)

Of course, all this is for the sake of advertising his work in a competitive marketplace which had seen previous biographies of Johnson, but there is more to it. Boswell expresses his opinion that if others had been as diligent in recording their memories of Johnson '*he* [Johnson] might have been almost entirely preserved', and he expresses his expectation that Johnson 'will be seen in this work more completely than any man who has ever yet lived' (1791/2008: 22). Biography thus understood *preserves the individuality of the deceased*; it is, Boswell claims, the continuation of the ancient 'pious office of erecting an honourable monument' to the hero's memory (1791/2008: 4). Boswell insists, however, that the *Life of Johnson* is 'not his panegyrick (...) but his Life; which, great and good as he was, must not be supposed to be entirely perfect' (1791/2008: 22). Boswell thus openly writes about Johnson's shortcomings, such as his excessive fear of death, which prompted the latter on at least one occasion to behave extremely uncivilly towards himself, and even indicates Johnson's sexual appetite for prostitutes. Boswell asks his readers, however, and this marks his biography as a modern enterprise, 'Let the question [of Johnson's sexual morality] be considered independent of moral and religious association' (1791/2008: 1376). The modernity of Boswell's *Life of Johnson*, in short, consists in its emphasis on the uniqueness of a person's life. References to the deceased's virtues, however, are still important markers in the narrative, and should underscore the greatness of the deceased.

Whilst the *Life of Johnson* is not itself a consolation, it nonetheless highlights the possibilities that an emphasis on biography provides for consolatory purposes; it foreshadows a strand of (auto)biographical memorialisation which aims to preserve the unique life of the deceased in its wholeness.[11] The

---

11 This emphasis on the wholeness of a unique life stands in stark contrast to a focus on deathbed conversations that had long roots in Christian spiritual literature and that did not cease to exist. Cf. for a specimen Hacker (1796), a tightly scripted attempt to draw spiritual lessons from the conversations and writings of famous dying men.

intensification of life suggested by this type of consolation is the realisation of biographical uniqueness. Another possibility is the self-consolation afforded by the autobiographical recording of the life of the bereaved. Grief memoirs, small-scale autobiographies that are limited to a rather short period following the loss, became successful on the book market around this time. Whilst it will take a long time for the genres of (auto)biography and grief memoir to develop the full potential foreshadowed by *Life of Johnson*, important elements are already visible in an early exemplar, Nehemiah Adams' anonymously published *Agnes and the Little Key* (1837/1863), which saw numerous editions in the USA and unauthorised reprints in England. The book begins with a sketch of the life and death of Adams' infant daughter Agnes. While much of the rest of the book is a recapitulation of traditional Christian consolatory motifs there are also new elements, such as detailed conversations about the loss that remind us rather of Boswell than of earlier consolatory texts; the minute autobiographical sketch of how the author was obsessed with the key to Agnes' coffin as a material focus of his thoughts about her; and the depiction of heaven as a domestic scene, elaborated by later authors.[12]

The fact that (auto)biographical memorialisation is still a fruitful strand of consolation should not cause us to underestimate its deep cultural roots. Whilst there is an abundance of recent popular grief memoirs (e.g. Didion 2005, van der Heijden 2011) and of funeral orations that have customarily had an important aspect of (auto)biographical memorialisation, arguably the best-known and certainly most-discussed exemplar of this strand of consolation remains C.S. Lewis's *A Grief Observed* (1961). This iconic text is a grief memoir, about a brief period following the loss of the author's wife. The text continues key elements already found in Boswell's *Life of Johnson*, in particular the wish to describe an individual life, without making it appear as the application of general moral or metaphysical lessons. Lewis, a devout Christian, reflects on the difference between his feelings, experiences and doubts on the one hand and what he is told (or what he previously thought) he should feel on the other.[13] He accepts the sadness that comes with the loss as the price of love. He cannot wish for a 'return to normal'.[14]

12 On the larger current of American consolation literature 1830–1880, see Douglas (1974).
13 This begins with the very first sentence 'No one ever told me that grief felt so like fear' (1961: 5) and continues throughout the book. For instance: 'They tell me H. is happy now, they tell me she is at peace. What makes them so sure of this?' (1961: 24) and 'I've read about that in books, but I never dreamed I should feel it myself' (1961: 46).
14 'It frightens me to think that a mere going back should even be possible. For this fate would seem to me the worst of all; to reach a state in which my years of love and marriage should appear in retrospect a charming episode – like a holiday – that had briefly interrupted my interminable life and returned me to normal, unchanged' (1961: 51).

All this is in keeping with (auto)biographical memorialisation as a consolatory strategy. It presupposes and at the same time supports the altered attitude towards grief already described: if remembrance preserves the deceased, grief must not be eradicated or suppressed. It makes far more sense to cultivate it on a lower level of intensity. Well-known historical phenomena such as the aestheticisation of death, the start of the custom of visiting graves (Ariès 1981), or Victorian mourning etiquette (Morley 1971), are all material and tangible expressions of this cultural development.

The most important challenges to (auto)biographical memorialisation came from two directions: first, the ascendancy of scientific world-views challenged belief in the uniqueness of human beings and the emphasis on the importance of individuality; death could thus become part of an economic calculus of efficiency. The new crematoria of the nineteenth century are expressions of this trend, even if they attempted to minimise the appearance of a breach with existing cultural formats by using Classical forms of architecture. Second, however, emerging large-scale conglomerates such as the nation and the modern state asserted their prevalence over, and control of, the individual. Typical expressions of this trend are war memorials since the French Revolution, with their celebration of the resilience of the community (Koselleck 2002). Against the background of mounting pressure against individuality, the unprecedented losses of life during WWI demonstrated the untenability of publicly cultivated grief and facilitated the development of a new consolatory strand.

3 Twentieth century/contemporary era: Professionalised consolation

Whilst (auto)biographical memorialisation not only allowed for attenuated residual grief but also aestheticised it, the highly elaborate cultivation of grief that was visible in the mourning rituals of the late nineteenth century must have seemed out of place and inapplicable in the face of the Great War's death toll. In this situation, there was a decided move towards publicly less disruptive and more efficient ways of dealing with grief. The public display of grief was reduced, and grief became increasingly private. Whilst appeals to the resilience of the consoland had been part and parcel of the earlier strands of consolation,[15] the emphasis on people's natural resilience when dealing with loss now became a dominant theme. In 'normal' cases, the bereaved could deal with loss in their own way, by drawing upon their spiritual resources and a social network of family and intimates. It became important to separate 'normal' losses from 'abnormal' or 'complicated' ones in which newly private grief was insufficient. A new class of consolers emerged, to act as gatekeepers between those categories and at the same time as providers of support: these were professional therapists with

---

15 See e.g. Seneca, *To Marcia on Consolation* 1.2–5 (Basore 1932: 3–7); Plutarch, *Consolation to His Wife* 609D (De Lacy and Einarson 1959: 589).

medical, psychoanalytical or psychological training, who offered a 'talking cure' to deal with abnormal grief.

However, we should not underestimate the long time it took for these developments to be fully realised; in its early days the therapeutic 'talking cure' was an elite phenomenon; it would take a long time, growing affluence, and higher levels of general education to make professionalised consolation an important factor in a democratic mass culture.

An early signpost of the trend towards professionalised consolation is Freud's *Mourning and Melancholia* (1917/1957), which has been regularly identified as the founding text of modern scientific studies of death (e.g. Walter 1996: 7; Klass et al. 1996: 5; Payne et al. 1999: 6; Walter 1999: 104; Archer 2008: 45). Freud disparages melancholy as a pathological reaction to loss, to be distinguished from non-pathological, normal grief or mourning (*Trauer*), which is a temporary affliction and can in principle be overcome. To guarantee the expertise of the professional consoler a scientific epistemological culture must be established, and *Mourning and Melancholia* delivers splendidly on this point. Whilst today's psychologists appear embarrassed by the mythological character of their predecessor's hypotheses, all essential ingredients of professional epistemology are present in Freud's text. Most importantly, Freud claims to provide a *general psychological theory* grounded in 'general observation' (1917/1957: 244). 'Grief work', 'libido', 'libidinal position', 'reality-testing', 'attachment', 'hypercathexis' and 'detachment' are concepts that *universally characterise* the mourning process, with therapy being needed when individuals cannot by themselves complete 'the work of mourning' to become 'free and uninhibited again' (1917/1957: 245). One can hardly overestimate the influence of those stipulations: the professional consoler claims to occupy a neutral, meta-level position. The therapist's theories and concepts unveil general truths about the client which hold good irrespective of the client's specific convictions. The therapist knows how the grief will unfold and what needs to be done. In the wake of Freud, later professional consolers and researchers will offer a plethora of terms to convey a sense of professional control – terms such as 'stages' (e.g. Kübler-Ross 1969), 'phases' (e.g. Bowlby 1980), 'tasks' (Worden 1983–2010), 'dual process' (Stroebe and Schut 1999) or 'scales' (Machin 2009). The therapist's meta-level position implies that the clients' world-views can be thematised and deployed in the therapeutic process. All this promises effectiveness and efficiency, since the therapist can screen, counsel and treat a wide variety of clients irrespective of background. It hardly needs emphasising that the new professional therapy shuns the language of consolation, which appears to be tied to an age of discredited religion and, perhaps even more importantly, to pre-scientific intervention. As such, consolation seems unable to offer the meta-level position occupied by the professional therapist and thus fails to convey the sense of energy, effectiveness and efficiency to which the professional consolers aspire. 'Coping', 'resilience', and 'process' become the new buzzwords.

The typical genre of this new type of consolation is the bereavement handbook, directed at the professional counsellor but also bought and read by a broader audience. Ideally, these handbooks are periodically updated to integrate new findings. The iconic example is J. William Worden's *Grief Counselling and Grief Therapy*, which has appeared in four major editions and numerous reprints between 1983 and 2010.[16] The four editions have seen the book double in length and reduce its dependency on Freud's theories. New emphases, such as the importance of maintaining 'continuing bonds' with the deceased or the 'Dual-Process Model of Grieving', are added without highlighting the impact of the alterations on the overall view of grief advocated in the book.

The bereaved have to complete four 'tasks' that are rephrased across the editions. In the fourth edition they are identified as 'to accept the reality of the loss', 'to process the pain of grief', 'to adjust to a world without the deceased' and 'to find an enduring connection with the deceased in the midst of embarking on a new life' (2010: 39–52). Dealing with grief here becomes a metaphor of life as a whole: its ideal of intensification is the successful completion of discrete tasks, the ticking of boxes – no small feat since conflicting demands must be balanced. The therapist can help, and the professional confidence exuding from the pages is remarkable. The therapist seems to know exactly what the mourning process ought to be like and can help the client to accomplish the tasks should they 'get stuck anywhere in the process' (2010: 172).

With professional therapy so clearly shunning the language of consolation, we need to reflect in more detail on the justifications for presenting this trend as a continuation of the consolatory tradition, or even as a 'third strand' of the work of consolation. In my view there are three important factors that allow us to understand professionalised therapy as such a strand: First, the goal of consolation, the attenuation of grief, is still present. Over the history of consolation there is considerable variation as to how far the attenuation should go (from eradication or suppression of grief to moderation and residual, lower-intensity grief), with twentieth-century and contemporary professionalised therapy falling comfortably between the extremes. Second, professionalised consolation continues to thematise world-views, the life of the deceased and its impact on the (auto)biography of the consoland. To refer to *Grief Counselling and Grief Therapy* again, Worden (2010: 73) notes that 'certain worldviews can serve a protective function by allowing individuals to incorporate a major tragedy into their belief system'; he prescribes close attention to the life of the deceased, their virtues and, as we can expect in the wake of (auto)biographical consolation, their less pleasant sides as well:

Considerable time is spent in the early sessions talking about the deceased, particularly about positive characteristics and qualities and pleasant

16   Dates of the UK editions.

activities that the survivor enjoyed with the deceased. Gradually begin to talk about some of the more mixed memories. The technique mentioned in chapter 4 can be useful here. 'What do you miss about him?' 'What don't you miss about him?' Finally, lead the person into a discussion of memories of hurt, anger, and disappointment.

(2010: 157–158)

Third, the claimed possession of a general theory with which to help the consoland is not entirely new: the pre-modern metaphysical and moral consolation also offered general theories (and was, in consequence, interrogated for their applicability to specific cases). The claim to a neutral stance on the part of the professionalised consoler is not entirely new, either; it is prefigured in the stance of the modern biographer, with Boswell as our prime example. However, what is new is the combination of the two: the professionalised consoler paradoxically claims to be in possession of a neutral standpoint that can be articulated over and above any world-view taken by the consoland; the professional world-view itself is not simply one among others.

It is this paradox that marks the main area of contention around the third strand of consolation: Is the scientific world-view of professional consolation sufficiently open to the life-experience of individuals and the world-views that they have? Klass (2014) inadvertently points to this area of contention when, in his rich description of a self-help group, he shows how the bereaved search for a community to validate their experiences and world-views. Apparently, the new professionalised consolation is insufficiently hospitable to experiences of continuing bonds with the deceased; it thus functions as a specific world-view that is perceived to be in conflict with, and to silence, other voices. The polemic against the Freudian concept of 'grief work', and especially the concept of 'closure', is merely one example of this trend (e.g. Gilbert 2006; Berns 2011). Sometimes, uneasiness with professionalised consolation is channelled in technologically new ways while reiterating older forms of consolation. The internet, for instance, has provided the bereaved with opportunities to create virtual memorials that can be continually updated, thus reflecting the will to cling to the dead. Such internet memorials utilise the earlier (auto)biographical memorialisation type of consolation. New ways of utilising the ashes of the deceased also appear to revisit earlier, Victorian models of memorialisation.

It would be too early, however, to claim that such uneasiness with the paradoxical language game of professionalised consolation marks its irrevocable decline, and for three reasons: First, professional academic grief specialists were themselves important and early critics of the Freudian concept of grief work (Wortman and Silver 1989; Stroebe et al. 1992; Klass et al. 1996). Second, grief theory building has proved to be quite resilient, with authors now claiming the integration of new ideas and hospitality to individual experiences (Walter 1996; Stroebe and Schut 1999; Machin 2009; Worden 2010). Third, professional expertise is not discredited lock, stock and barrel; it still carries weight with the general public, as is evidenced by advertisements of professional training by many authors of experience-based consolatory books (e.g. Kachler 2012). Although it is hard to make predictions, the strand of professionalised consolation currently seems resilient.

The Four-Axis Model of Consolation

Our brief survey of the history of consolation discloses considerable variation across the three strands, but it has also made visible a certain continuity of issues and approaches over time. Recurring themes are the goals of attenuating grief and of increasing resilience, the negative view of unchecked grief, the appeal to the resilience of the survivor(s), the importance of world-views and the narrative preservation of the life of the deceased.

In the construction of a conceptual framework of consolation we need to aim at a reflective equilibrium between the systematic ambition of providing a general theory and the interpretation of specific cases (Rawls 2005). The framework should have the overarching quality of integrating and contextualising the previous proposals and the strands in the history of consolation while demonstrating that not all consolation is 'the same'. Whereas the goals are constitutive for consolation, the other recurring themes can best be thought of as four 'variables' that take significant (positive) values over time. Together, goals and variables form the *conceptual core* of consolation. The best means of depicting the conceptual core of consolation appears to be a *radar chart*, with the four variables as *axes*. One of the advantages of radar charts is that they can easily be extended. One could subdivide themes or variables to measure the relative importance of different ideas belonging to one recurring theme. For instance, the theme of the importance of world-views insofar as they offer 'healing views of death' (i.e., give death a place in the greater scheme of things) aggregates different motifs such as death as sleep, the cycle of nature, emphasis on a personal afterlife and emphasis on a theistic God rewarding good lives, which could be depicted by different axes, should such differentiation be felt useful. One could also add new axes in order to chart further differences between types of consolation not thematised in the conceptual core – for instance, the degree to which they acknowledge gender differences.[17] It is, however, my contention that the four axes here identified form the conceptual core of consolation. New axes can be valuable, although they diffuse the conceptual core that identifies recurring themes.

The conceptual core of consolation can be outlined as follows: *Consolation is a family of practices and interventions (comforting gestures, rituals, oral or written communication, etc.) that aim to reduce distress produced by the experience of the limitations of human existence, to restore as much as possible*

---

17 Such acknowledgments sometimes are, but need not be, part of emphasis on the resilience of the survivor. See e.g. Seneca, *To Marcia on Consolation* 16.1–4 (Basore 1932: 49–51).

normal functioning, and to foster resilience in the face of past as well as (unspecified) future losses. Consolation functions along *four axes*:

(1) Consolation is about the regulation of grief. It tends to presuppose a negative view either of grief in general or of excessive, unchecked grief; it sets against such unchecked grief an *ideal of acceptable grief*: a form of grief that all can live with. The regulation or 'policing' (Walter 1999) of grief thus prescribes a relationship between the individual and the community. In written consolations, the relationship is sometimes thematised explicitly, in higher-order reflections: it is pointed out, for instance, that grief is socially constructed and therefore malleable: grief is, or can eventually be, in the control of the bereaved.[18]

(2) Consolation attempts to *rescript the survivors' lives by appealing to their resilience*: it nudges them away from potentially self-destructive and towards constructive attitudes and forms of behaviour. In written consolations, we frequently find that important sources of inner strength are identified: the bereaved is stronger than he/she thinks, has coped with earlier losses and challenges in an admirable way, and is part of a community that is resilient.[19]

(3) Consolation offers a *therapeutic account of death*: it offers (ingredients for) *a world-view* geared at showing how death can be accepted in spite of all the grief it causes. In written consolations, *healing narratives* are related. To name a few obvious examples: death is part of nature; death is like sleep; or death is translocation to a homelike scene. The function of these healing narratives is to provide a general metaphysical framework in which to locate the specific death. Although they are told on the occasion of a specific death, they are in principle applicable to others. Whilst such narratives provide a high level of abstraction, the next theme is about the concrete life of the deceased.

(4) Consolation tries to *'rewrite' the life of the deceased and thus to preserve it in a narrative*: whereas grief emphasises the incompleteness of the life lost and all that might have been, consolation counteracts such thoughts and feelings by emphasising the senses in which the deceased's life has been fulfilled and is complete. This should not be taken to mean that the deceased is 'done with' or irrelevant for the present. On the contrary, trying to establish the true nature and legacy of the deceased serves to heighten their reality; to show how the deceased is still relevant to our lives and is somehow 'still there'. Recent 'continuing bonds' literature has reiterated the frequency with which the deceased remain present in their survivors' lives. In written

18 See e.g. Plutarch, *Consolation to His Wife* 611A–B (De Lacy and Einarson 1959: 599). The negative view of grief seems to be contradicted by sources of memorialising consolation that reject an end to grief. My reply is that the contradiction is only apparent; as we can see in Lewis (1961); it is not the unchecked, excessive grief that is continued; it is a residual, reduced grief, much like melancholy in Goethe (1774). Second, the need not to end grief does not imply a favourable interpretation of grief; rather, grief is seen as the lesser evil compared to forgetting the deceased that is associated with the end of grief.

19 See e.g. Plutarch, *Consolation to His Wife* 609C–E (De Lacy and Einarson 1959: 589–591).

consolations the legacy of the deceased is regularly emphasised: what was central to the deceased (work, family commitments, spare-time activities) lives on, sometimes also in large-scale imagined communities such as the nation. This axis of consolation provides a clear *(auto)biographical focus*. However, it should be acknowledged that the ways of writing (auto)biographies have undergone marked changes from classical antiquity to today. Whilst a key ingredient of premodern metaphysical and moral consolation continues with an emphasis on the virtues of the deceased, there is today, in the wake of modern biography, more scope for recognising the deceased's typical (and sometimes endearing) vices. A radar chart depiction might look as per Figure 1.1.

Further, it is important to distinguish between consolation in the above narrow or literal sense and an extended or metaphorical use of 'consolation'. The core of the concept of consolation is premised on an actor purposefully using practices and interventions to reduce distress and increase resilience. However, we often use the concept of consolation *in an extended or metaphorical sense*, without being committed to an actor purposefully attempting to console. We find such an extended or metaphorical use in phrases of the type 'S experiences/finds consolation in x'; for instance: 'He finds consolation in the forests around his home town where he used to take long walks with his wife.' The continuation of a joint important activity or the experience of a specific landscape are here presented as if they were agents; it is *as if* a specific

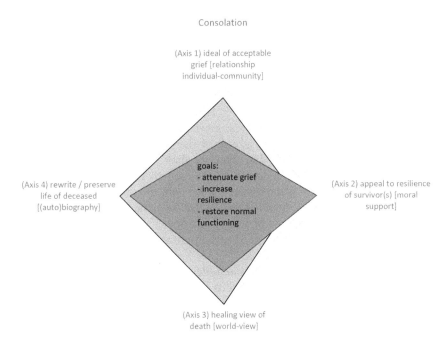

Figure 1.1 The Four-Axis Model of Consolation

landscape had tried to reduce our distress with its splendour, or *as if* an activity we undertake was purposefully increasing our resilience.

Regardless of the extended or metaphorical ('as if') use of the phrase, the conceptual core and the four themes of consolation as outlined above provide hermeneutical tools to interpret such utterances: Is the continuation of the former joint activity what is emphasised here? Is it important because it symbolises a re-discovery of the control (Axis 1), or important sources of inner strength (Axis 2), and/or because it continues and thus validates something that was important to the deceased (Axis 4)? Is the experience of the landscape emphasised in the utterance? Is it important because it offers a therapeutic account of death, such as death as part of nature or the continual regeneration of nature as a symbol for life (Axis 3)?

Application

If the conceptual framework outlined thus far is adequate, it then has to prove its value as a hermeneutical tool. It should suggest stimulating interpretations of phenomena and generate relevant questions. Whilst it is outside the scope of the present chapter to enter into detailed analyses of specific cases, I wish to suggest lines of inquiry in respect of typical spatial constellations of grief and consolation. In so doing, I use Maddrell's differentiation of three categories of grief/consolation spaces (Maddrell 2016, and Chapter 2 in this volume), and discuss briefly a phenomenon in each category. Maddrell distinguishes between (1) physical or material spaces, (2) embodied-psychological spaces, and (3) virtual spaces.

1 Physical or material spaces

Physical or material spaces can range in scale, from small artefacts and public formal or informal memorials to evocative landscapes. A key set among them 'are those physical spaces of burial, cremation and memorialisation' (Maddrell 2016: 174). To focus on cemeteries and grave-markers, ancient cemeteries such as the Athenian Keirameikos already show the practice of producing inscriptions that name the deceased, and there we find a memorialisation of sorts. However, with the rise of modern individualised biographical memorialising consolation, this function is intensified and brought to a new level. This results in a two-fold development: first, not to preserve the names of the deceased is to heighten anxiety, as the spread of war cemeteries shows. By symbolically particularising a single, anonymous dead soldier, graves of The Unknown Warrior address that anxiety (see Figure 1.2).[20]

The existence of civilian unnamed dead is felt as a moral scandal in modern affluent societies, and city councils tend to offer symbolic memorials for those buried or cremated under their responsibility

---

20 Copyright: Dean and Chapter of Westminster. Note the use of 2 Chronicles 24:16 (Jehoiada) to attribute virtues to the fallen (see below).

Figure 1.2 Grave of the Unknown Warrior (Westminster Abbey, London, 1920)

(see Figure 1.3).[21] Second, there is growing dissatisfaction with the limited memorialisation available in highly regulated cemeteries, which has led some observers to hypothesise the 'end of the traditional cemetery' (Sörries 2011). The obvious consequence of this development is that cemeteries are supplemented by other forms of memorialisation (such as memoirs), and with the advent of new technology moved to virtual spaces (see below). In extreme cases which dovetail with the increasing privacy for grief over the twentieth century, traditional cemeteries are bypassed altogether (through mobility of cremated remains and natural burials), creating private monopolies of memorialisation.

Whilst the distinction between three strands of consolation thus permits us to interpret important changes in burial culture, the conceptual framework of consolation allows us also to identify key elements with undiminished importance over time.

To single out one aspect, rewriting the life of the deceased (Axis 4) is an obvious cultural constant. Already in the Ancient Greco-Roman world, grave-markers and monuments identify the deceased person's important

21 See also Sörries (2011), who refers to a field for anonymous burials in Hamburg-Öjendorf, on which stones are placed with the inscriptions of generic roles, such as 'Friend', and to Sybille Löw's installation *Stiller Abtrag*.

Figure 1.3 Memorial to those who died without friends or family (Selwerderhof, Groningen)

virtues. There is probably no grander example of this than the Celsus Library in Ephesus (second century CE). Tiberius Julius Aquila intended to build a grand memorial for his father Celsus, but the city council demanded that the building would also function as a library. The monumental façade contained four statues evoking the virtues of Celsus (e.g. 'Celsus' wisdom', see Figure 1.4).

The evocation of the virtues of the deceased is continued to this day. Contemporary grave markers can contain allusions to the triadic Christian virtues faith, hope and love, but frequently we find more personal ascriptions. The details below (Figure 1.5, from a cemetery in the north of the Netherlands) show the humour, courage and loving care of the deceased (Jedan 2017b). As we will see below, contemporary internet memorials elaborate on the virtues of the deceased, unconstrained by the material and cemetery regulations.

As a rule, it is important not to forget the limitations imposed by specific spatial arrangements on consolation along Axis 4. War memorials, for instance, tend to prioritise the community in whose service soldiers left their lives, and Green burials certainly tend to offer little in terms of Axis 4. The fact, however, that Green burial site personnel often need to discretely remove illegitimate grave markers bears witness to the importance of Axis 4.

Figure 1.4 Detail of the façade of the Celsus Library, Ephesus

2 *Embodied-psychological spaces*

The inclusion of embodied-psychological spaces is an important aspect of Maddrell's conceptual framework. She explains that 'Valuable as studies of memorial places and landscapes are, it is necessary to move beyond them in order to understand the full range of spaces significant in bereavement, grief and mourning. (…) mourners are embodied and it is central to the framework presented here to acknowledge the embodiment of grief as a space of experience, practice, performance and trace' (2016: 176).

Psychosomatic responses to loss and psychosomatic aspects of consolation are *key* phenomena in this category. The psychosomatic nature of grief and consolation is widely recognised in written consolations. It ranges from ancient sources distinguishing between unchecked and controlled grief behaviours to contemporary psychological handbooks contrasting normal and abnormal grief behaviours (e.g. Worden 2010: Ch. 5).

The emphases on biography and on autobiographical scrutiny make modern grief memoirs particularly interesting sources. In *A Grief Observed* (1961), Lewis juxtaposes grief and a feeling of consolation (although without actually using the

Figure 1.5 Details of contemporary grave inscriptions

term 'consolation' he describes it as 'the lifting of the sorrow'), and points repeatedly to their psychosomatic nature. For instance, at the previously quoted start of *A Grief Observed* he likens grief to fear ('No one ever told me that grief felt so like fear' – 1961: 5) and immediately inscribes his physical reactions: 'The same fluttering in the stomach, the same restlessness, the yawning. I keep on swallowing' (1961: 5). When Lewis reports how, to his surprise, he is feeling better ('Something quite unexpected has happened' – 1961: 38), he likewise notes the psychosomatic connection: 'my heart was lighter than it had been for many weeks. For one thing I suppose I am recovering physically from a good deal of mere exhaustion (…) a sounder night's sleep' (1961: 39). For our purposes it is particularly interesting to see how this experience of consolation is connected to memorialisation and to the fourth axis of consolation: 'And suddenly at the very moment when, so far, I mourned H. least, I remembered her best. Indeed it was something (almost) better than memory; an instantaneous, unanswerable impression. To say it was like a meeting would be going too far. Yet there was that in it which tempts one to use those words' (1961: 39).

3 Virtual spaces

This category 'encompasses all non-material spaces of interaction, practice and performance', with the online memorial as an increasingly significant example (Maddrell 2016: 178). I propose to look at the internet memorial of

Nicki-Sophie (d. 2004), which has been described as 'typical' by an earlier researcher (Gebert 2009).[22]

It is striking that the initial page is titled 'Nicki's Home'. Visitors are greeted by images of Nicki-Sophie and are encouraged to go in. At a mouse-click they enter a memorial space made homely by background music. They have crossed a virtual threshold, marking a certain intimacy (which can be expected in an age of private grief) against the background of the global availability of such a site. Among the contents of the site, much room is given to the reconstruction of the road accident that killed Nicki-Sophie; it firmly places the guilt with a certain 'R.L.', who drove the car that knocked her down. It presents the legal aftermath of the tragedy and its press coverage. Apart from the technical possibilities afforded by the internet age, all this is consonant with the modern memorialising strand of consolation already visible in Boswell's *Life of Johnson*: it is the attempt to establish the facts about a life, or at any rate to present the compiler's own perspective on that life. It would be a mistake to interpret such memorials stereotypically as a 'postmodern' shift of authority away from professional consolers. One heading is devoted to 'grief counselling' (*Trauerbegleitung*), where the parents thank Mechthild Ludwig-Mayer, a professional grief counsellor and funeral speaker. Some of the pictorial material, such as the image of Nicki and her dog on a cloud utilise motifs of heaven as a home, which became established in the era of modern memorialising consolation (Axis 3). Images of Nicki-Sophie's grave, replete with the figure of an angel, are presented under the heading 'Nicki's Garden', underpinning the interpretation of the afterlife as a domestic scene.

In the context of physical spaces of consolation, we have seen that physical grave-markers often display messages in line with Axis 4: they rewrite the life of the deceased by emphasising their virtues. In this sense also, Nicki-Sophie's life is rewritten: small stories are added to the overall picture of a lovely girl ('a princess'). Those stories abound with terms indicative of virtue, such as: 'attentive', 'intelligent', 'always the brightest child in her class', 'never arrogant', 'always willing to help others', 'excellent physical prowess' (inline-skating, cheerleading), 'modest' and 'great sense of justice'.

Conclusion

We have seen that the conceptual framework offers strong possibilities for new interpretations of familiar phenomena. In particular, the framework allows us to locate consolatory phenomena historically and to appreciate, beside the changes, the continuities in consolatory practices. The framework thus counteracts tendencies to overemphasise the novelty of recent developments. Practices such as the memorialisation of the deceased person's biography are well-established and by no means the unique expression of a contemporary way of life.

But, for all its relevance as interpretative tool, does the conceptual framework generate relevant research questions? In my view, the particular value of

---

22 http://www.nicki-sophie.de/index.html (last accessed 12 July 2018).

the conceptual framework lies in its capacity to cut across both different eras and also categories of physical, embodied-psychological and virtual phenomena, thereby suggesting interesting comparative questions.

To mention a few of them: In the foregoing analysis, it was suggested that internet memorials, while not subject to restrictions of material and space in the same way as physical memorials, can elaborate on the virtues ascribed to the dead (Axis 4). Are there also differences in the virtues attributed? If so, how is the creation of intimacy in virtual spaces connected to the virtues being attributed? What strategies permit us to create intimacy across the three spatialities? Even if the attribution of virtues remains constant, have the virtues customarily ascribed to the deceased person changed over time? What are the most important virtues utilised nowadays across the three spatial categories? To what extent do they reflect broader cultural changes? Have Axis 3 narratives changed over time? It is clear that ideas of the afterlife as a domestic scene, developed in the nineteenth century, are still an important component of consolation today. Are there also new motifs in evidence that fulfil the same function in a new way (e.g. by response by technological developments)?

A final word on the likely limitations of the conceptual framework: The general concept has been abstracted from a broad array of sources that have become important in Western cultural histories. Stating that sources *have become important* in the West is not to endorse the oversimplified claim that they *are* 'Western'. We are only beginning to appreciate the extent to which Greco-Roman Antiquity was outward-looking and able to integrate cultural influences from far afield. Nevertheless, there can be no *a priori* assurance that, for instance, such a framework fits indigenous consolatory practices in the Global South.

At the same time, it is worth bearing in mind that whilst consolation is an important aspect of our dealings with death, it is by no means the only aspect. Practical considerations may play an important role. In the Netherlands, for instance, with respect to Green burials, perpetual grave rights and lack of any long-term care requirement for a grave appear to be far more important attractions than comprehensive ecological word-views (Herder 2016). With some sites, the intention of their creators might be a pointed *denial* of consolation: Peter Eisenman's Berlin Holocaust Memorial is a well-known example, and arguably a number of war memorials can better be understood as demonstrations against war than as consolations. Further research is needed to test the present framework's explanatory reach, as well as consolation's relationship to other aspects of our dealings with death.[23]

23 I am grateful to Tony Walter and Eric Venbrux for commenting on a draft of the chapter and suggesting further literature. Research for this chapter has benefitted from a residential fellowship at the Netherlands Institute for Advanced Study in the Humanities and the Social Sciences (Royal Netherlands Academy of Arts and Sciences), NIAS-KNAW, in 2016/17.

References

[Adams, N.] 1837/1863. *Agnes and the Little Key: Or, Bereaved Parents Instructed and Comforted. By her father.* Eighth revised edition, Boston: Ticknor and Fields.

Archer, J. 2008. 'Theories of grief: past, present, and future perspectives', in *Handbook of Bereavement Research and Practice: Advances in Theory and Intervention*, edited by M. Stroebe, R. O. Hansson, H. Schut and W. Stroebe. Washington, DC: American Psychological Association, 45–65.

Ariès, Ph. 1981. *The Hour of Our Death: The Classic History of Western Attitudes toward Death over the Last One Thousand Years.* Translated by Helen Weaver. New York: Vintage.

Attig, T. 2011. *How We Grieve: Relearning the World.* Second edition. New York: Oxford University Press.

Babbitt, F. C., ed. and trans. 1928. *Plutarch, Moralia, Volume II: How to Profit by One's Enemies. On Having Many Friends. Chance. Virtue and Vice. Letter of Condolence to Apollonius. Advice About Keeping Well. Advice to Bride and Groom. The Dinner of the Seven Wise Men. Superstition.* Cambridge, MA: Harvard University Press.

Basore. J. W., ed. and trans. 1932. *Seneca, Moral Essays, Volume II: De Consolatione ad Marciam. De Vita Beata. De Otio. De Tranquillitate Animi. De Brevitate Vitae. De Consolatione ad Polybium. De Consolatione ad Helviam.* Cambridge, MA: Harvard University Press.

Berns, N. 2011. *Closure: The Rush to End Grief and What It Costs Us.* Philadelphia, PA: Temple University Press.

Boswell, J. 1791/2008. *Life of Johnson.* Oxford: Oxford University Press.

Bowlby, J. 1980. *Attachment and Loss. Volume III. Loss, Sadness and Depression.* New York: Basic Books.

Burrows, M. S. 1991. *Jean Gerson and De Consolatione Theologiae (1418): The Consolaiton of A Biblical and Reforming Theology for a Disordered Age.* Tübingen: Mohr Siebeck.

Choron, J. 1963. *Death and Western Thought.* New York and London: Macmillan.

Davies, D. J. 2002. *Death, Ritual and Belief: The Rhetoric of Funerary Rites.* Second edition. London and New York: Continuum.

Davies, D. J. 2005. *A Brief History of Death.* Malden, MA: Blackwell.

Davies, D. J. 2008. *The Theology of Death.* London and New York: T&T Clark.

De Lacy, Ph. H., Einarson, B., eds. and trans. 1959. *Plutarch, Moralia, Volume VII: On Love of Wealth. On Compliancy. On Envy and Hate. On Praising Oneself Inoffensively. On the Delays of the Divine Vengeance. On Fate. On the Sign of Socrates. On Exile. Consolation to His Wife.* Cambridge, MA: Harvard University Press.

Didion, J. 2005. *The Year of Magical Thinking.* New York: Knopf.

Douglas, A. 1974. Heaven our home: consolation literature in the northern United States, 1830–1880. *American Quarterly* 26/5, 496–515.

Erasmus of Rotterdam, D. 1539. *De praeparatione ad mortem.* Cologne: Eucharius Cerniconus.

Feuerbach, L. 1830/1960. Gedanken über Tod und Unsterblichkeit, in Feuerbach, L. *Sämtliche Werke. Vol. 1.* Stuttgart-Bad Cannstadt: Frommann Holzboog.

Fowler, H. N., ed. and trans. 1936. *Plutarch. Moralia, Volume X: Love Stories. That a Philosopher Ought to Converse Especially With Men in Power. To an Uneducated Ruler. Whether an Old Man Should Engage in Public Affairs. Precepts of Statecraft.*

On Monarchy, Democracy, and Oligarchy. That We Ought Not to Borrow. Lives of the Ten Orators. Summary of a Comparison Between Aristophanes and Menander. Cambridge, MA: Harvard University Press.

Freud, S. 1917/1957. Mourning and melancholia, in *The Standard Edition of the Complete Psychological Works of Sigmund Freud. Volume XIV.* London: The Hogarth Press, 243–258.

Gebert, K. 2009. *Catarina unvergessen: Erinnerungskultur im Internetzeitalter.* Marburg: Tectum.

Gerson, J. *c.*1414/*c.*1477. *Opus tripartitum de praeceptis decalogi, de confessione et de arte moriendi.* Deventer: Pafraet.

Gilbert, S. M. 2006. *Death's Door: Modern Dying and the Ways We Grieve.* New York: Norton.

Goethe, J.-W. 1774/1985. *Die Leiden des jungen Werthers.* Stuttgart: Reclam.

Hacker, J. B. N. 1796. *Thanathologie oder Denkwürdigkeiten aus dem Gebiete der Gräber: Ein unterhaltendes Lesebuch für Kranke und Sterbende.* Leipzig: Rein.

Hegel, G. W. F. 1796–97/1971. Das älteste Systemprogramm des deutschen Idealismus, in Hegel, G. W. F., *Werke 1: Frühe Schriften,* 234–236.

van der Heijden, A. F. Th. 2011. *Tonio: Een requiemroman.* Amsterdam: De Bezige Bij.

Herder, M. 2016. *Greening Death: Issues of Choice, Control, Meaning and Ritualization surrounding Natural Burial.* Unpublished MA thesis: University of Groningen.

Holloway, Paul A. 2001. *Consolation in Philippians: Philosophical Sources and Rhetorical Strategy.* Cambridge: Cambridge University Press.

Jedan, C. 2014a. Troost door argumenten: herwaardering van een filosofische en christelijke traditie. *Nederlands Theologisch Tijdschrift* 68/1&2, 7–22.

Jedan, C. 2014b. Cruciale teksten: de Grieks-Romeinse consolatio. *Nederlands Theologisch Tijdschrift* 68/1&2, 165–173.

Jedan, C. 2017a. The rapprochement of religion and philosophy in ancient consolation: Seneca, Paul, and beyond, in *Religio-Philosophical Discourses in the Mediterranean World: From Plato, through Jesus, to Late Antiquity,* edited by A. K. Petersen and G. H. van Kooten. Brill: Leiden, 159–184.

Jedan, C. 2017b. *Een voltooid leven: Over troost en de intelligentie van religie.* Groningen: University of Groningen.

Johann, H.-Th. 1968. *Trauer und Trost: Eine quellen- und strukturanalytische Untersuchung der philosophischen Trostschriften über den Tod.* Munich: Fink.

Kachler, R. 2012. *Meine Trauer wird dich finden! Ein neuer Ansatz in der Trauerarbeit.* Stuttgart: Kreuz.

Kassel, R. 1958. *Untersuchungen zur griechischen und römischen Konsolationsliteratur.* Munich: Beck.

Kauffman, J., ed. 2002. *Loss of the Assumptive World: A Theory of Traumatic Loss.* New York: Routledge.

King, J. E., trans. 1927. *Cicero, Tusculan Disputations.* Cambridge, MA: Harvard University Press.

Klass, D. 1993. Solace and immortality: bereaved parents' continuing bonds with their children. *Death Studies* 17/4, 343–368.

Klass, D. 2006. Grief, religion, and spirituality, in *Death and Religion in a Changing World,* edited by K. Garces-Foley. Armonk, NY: Sharpe, 283–304.

Klass, D. 2013. Religion and spirituality in loss, grief, and mourning, in *Handbook of Thanatology,* edited by D. K. Meagher and D. E. Balk. Second edition. New York and London: Routledge.

Klass, D. 2014. Grief, consolation, and religions: a conceptual framework. *Omega* 69, 1–18.

Klass, D. (unpublished). The nature of religious consolation for the bereaved. Unpublished draft. Available at: http://www.academia.edu/2247617/The_Nature_of_Religious_Consolation_for_the_Bereaved. Accessed 12 July 2018.

Klass, D., Silverman, Ph. R. and Nickman, S. L., eds. 1996. *Continuing Bonds: New Understandings of Grief.* London and New York: Routledge.

Koselleck, R. 2002. War memorials: identity formations of the survivors, in Koselleck, R., *The Practice of Conceptual History.* Stanford: Stanford University Press, 285–326.

Kübler-Ross, E. 1969. *On Death and Dying: What the Dying Have to Teach Doctors, Nurses, Clergy, and Their Own Families.* New York: Scribner/Macmillan.

Kunkel, A. D. and Dennis, M. R. 2003. Grief consolation in eulogy rhetoric: an integrative framework. *Death Studies* 27, 1–38.

Lewis, C. S. 1961. *A Grief Observed.* London: Faber and Faber.

Luther, M. 1519/1990. Ein Sermon von der Bereitung zum Sterben, in M. Luther, *Schriften, Volume 2.* Frankfurt: Insel, 15–34.

Luther, M. 1520/1891. Vierzehn Trostmittel für Mühselige und Beladene, in *Luthers Werke, Vol 7*, edited and translated by G. Buchwald et al. Braunschweig: Schwetschke, 5–60.

Machin, L. 2009. *Working with Loss and Grief: A New Model for Practitioners.* Los Angeles, CA, and London: SAGE.

Maddrell, A. 2016. Mapping grief: a conceptual framework for understanding the spatial dimensions of bereavement, mourning and remembrance. *Social & Cultural Geography* 17/2, 166–188.

Mette, H.-J. 1984. Zwei Akademiker heute: Krantor von Soloi und Arkesilaos von Pitane. *Lustrum* 26, 7–94.

Miller, C. L., trans. 1998. *Jean Gerson, The Consolation of Theology.* New York: Abaris.

Morley, J. 1971. *Death, Heaven and the Victorians.* London: Studio Vista.

Neimeyer, R. A., ed. 2001. *Meaning Reconstruction and the Experience of Loss.* Washington, DC: American Psychological Association.

Nietzsche, F. 1889/1988. Götzen-Dämmerung, in *Friedrich Nietzsche: Kritische Studienausgabe. Vol 6*, edited by G. Colli, and M. Montinari. Munich, Berlin and New York: dtv and de Gruyter, 55–161.

Norberg, A., Bergsten, M. and Lundman, B. 2001. A model of consolation. *Nursing Ethics* 8/6, 544–553.

O'Connor, C. 1966. *The Art of Dying Well: The Development of the Ars Moriendi.* New York: AMS.

Parkes, C. M. and Prigerson, H. G. 2010. *Bereavement: Studies of Grief in Adult Life.* 4th edition. London: Penguin.

Payne, S., Horn, S. and Relf, M., eds. 1999. *Loss and Bereavement.* Buckingham and Philadelphia, PA: Open University Press.

Quint, J., ed. and trans. 1979. *Meister Eckehart, Deutsche Predigten und Traktate.* Diogenes: Zurich.

Rawls, J. 2005. *Political Liberalism.* Expanded edition. New York: Columbia University Press.

Schleiermacher, F. 1799/1969. *Über die Religion: Reden an die Gebildeten unter ihren Verächtern.* Stuttgart: Reclam.

Smith, A. 1759/2000. *The Theory of Moral Sentiments.* New York: Prometheus.

Sörries, R. 2011. *Ruhe sanft: Kulturgeschichte des Friedhofs.* Kevelaer: Butzon & Bercker.

Stammkötter, F.-B. 1998. 'Trost', in *Historisches Wörterbuch der Philosophie 10*, 1523–1527.

Stroebe, M. and Schut, H. 1999. The dual process model of coping with bereavement: rationale and description. *Death Studies* 23/3, 197–224.

Stroebe, M., Gergen, M. M., Gergen, K. J. and Stroebe, W. 1992. Broken hearts or broken bonds: love and death in historical perspective. *American Psychologist* 47/10, 1205–1212.

Tester, S. J., trans. 1973. *Boethius, The Theological Tractates: The Consolation of Philosophy.* Cambridge, MA and London: Harvard University Press.

Walter, T. 1994. *The Revival of Death.* London: Routledge.

Walter, T. 1996. A new model of grief: bereavement and biography. *Mortality* 1/1, 7–25.

Walter, T. 1997. 'Letting go and keeping hold: a reply to Stroebe'. *Mortality* 2/3, 263–266.

Walter, T. 1999. *On Bereavement: The Culture of Grief.* Buckingham and Philadelphia: Open University Press.

Weyhofen, H. Th. 1983. *Trost: Modelle des religiösen und philosophischen Trostes und ihre Beurteilung durch die Religionskritik.* Frankfurt and Bern: Lang.

Wijdeveld, G., trans. 2003. *Thomas a Kempis, De navolging van Christus naar de Brusselse autograaf.* Fifth edition. Kapellen and Kampen: Pelckmans and Ten Have.

Worden, J. W. 1983. *Grief Counselling and Grief Therapy.* London: Tavistock.

Worden, J. W. 1991. *Grief Counselling and Grief Therapy: A Handbook for the Mental Health Practitioner.* London: Routledge.

Worden, J. W. 2003. *Grief Counselling and Grief Therapy: A Handbook for the Mental Health Practitioner.* Third edition. London: Brunner-Routledge.

Worden, J. W. 2010. *Grief Counselling and Grief Therapy: A Handbook for the Mental Health Practitioner.* Fourth edition. London: Routledge.

Wortman, C. B. and Silver, R. C. 1989. The myths of coping with loss. *Journal of Consulting and Clinical Psychology* 57/3, 349–357.

http://www.nicki-sophie.de/index.html (last accessed 12 July 2018).

2 Bittersweet

Mapping grief and consolation through the lens of deceased organ donation

Avril Maddrell

Introduction

> Recently visiting my Father in hospital where he was recuperating after a fall, I was moved in various ways: by his injuries, by the care provided by dedicated NHS staff, and by a memorial garden of remembrance enfolded within the structure of the hospital buildings. It was a garden dedicated to the memory of the deceased whose organs had been donated in the hope of restoring health – life even – to others. For a moment I was caught up in its affective atmosphere, sensitive to who was represented in this place, and the ever-recalibrating confluence of loss, love and pride held within for the bereaved.

This chapter begins with a discussion of the relationship between loss and consolation and how overlapping experience of both can be apprehended and better understood through attention to these being embedded in the intersecting spatialities of material places, embodied experiences and virtual arenas, as represented in the 'Mapping Grief and Consolation' framework. The second part of the chapter reviews discourses of consolation found in international studies of deceased organ donation; and the final section explores expressions and accounts of loss and consolation through the lens of a case study of a local organ donation activist group, highlighting the co-constitutive roles of physical memorial garden, embodied emotional-affective experience and social media platforms.

As outlined in the Introduction to this volume, consolation has become a devalued term in Western culture. The notion of 'consolation prize' dominates popular usage, rendering consolation as something of limited value or prestige, a sop given to soften the blow of failure, or, if not failure as such, of not winning. Much of the meaning and experience of consolation is lost through this narrowed vernacular interpretation. Grounded in my own experience of loss, and within my own work exploring the potential geographies of death, bereavement, mourning and remembrance, understanding loss and consolation have gone hand in hand: they are relational (see Maddrell 2009a). This chapter affords an opportunity to bring to the fore the place of consolation in relation to understanding the geographies of loss and remembrance.

Consolation for the bereaved is often associated with the comfort derived from faith and associated ritual (Klass 2014), and of cherished memories. Some find solace in active commemoration (Santino 2006), through the act of *doing*, e.g. lobbying or fundraising, which in Western society has been associated with masculine instrumental approaches to mourning (Martin and Doka 2000). However, this gendered interpretation has been challenged both by evidence of bereaved women's roles in such public activities and other undertakings such as maintenance, familial care and volunteering (Maddrell 2013). These activities can serve as mechanisms for side-stepping the full emotional force of bereavement, or, more commonly, are interleaved with more obvious emotion-focused experiences of mourning over time (see Stroebe and Schut [1999] on the psychological oscillation between loss and restoration in bereavement).

Time is often privileged in scholarship on mourning and living with loss (Maddrell 2016), but time and space intersect in the shifting place-temporalities of grief and consolation. For some, comfort is found in the spatial-temporal strategy of forgetting the dead who are located in the margins of place and memory (Robinson 2010); but for many, *remembering* is a subconscious psycho-biological response, and/or an emotional necessity. Conscious acts of remembering include acts of naming and memorialising the deceased, through artefacts, narratives and practices. Comfort can be found through the storying of the deceased (Walter 1996; Valentine 2008), beliefs about their continued afterlife (Maddrell 2009a, 2009b, 2011; Klass 2014), and acts of devotion and care manifest at and through sites of memorialisation (Maddrell 2013), each of which have their own geographies. Consolation can be found in geography, particularly the geographies of place, locations where geographical attachment and identity coalesce with the personal geographies of individual and shared lives, with associated memories and meanings. This is reflected in sites of bodily disposal, spiritual practice, and increasingly, in the West, through wider practices of private memorialisation in public spaces (e.g. in post-secular Britain). However, when considering the geographies of loss and consolation, other spaces need to be highlighted as well, particularly the space of the body and virtual arenas of activity and meaning-making.

This following discussion of consolation in bereavement, draws and builds on my previous reflections on geographies of loss and remembrance. It outlines a conceptual framework for understanding the *spatial* dimensions of experiences and practices associated with loss and remembrance, and how these in turn can be experienced – and even sought or curated – as spaces and practices of *consolation*. Consolation can be found in a variety of forms that are woven *through and beyond* these spaces animated by loss, such as sites associated with memory and/or continuing bonds. Expressing and practising relation to the deceased through continuing bonds can be manifest in various ways, such as through embodied practices such as talking to or praying for the deceased (see Klass 2014), caring for the deceased vicariously through the physical care and maintenance devoted to a memorial or shrine (cleaning, painting, gardening), or through acts in the name of the deceased such as fundraising or lobbying, which keeps their name in circulation and an ongoing 'life' for the named person (Maddrell 2013). This sense of liveliness in death (Stevenson et al. 2016) is sought-found in new activities or mobilities fulfilled on behalf of, or fulfilling the wishes of, the deceased: 'What they always wanted to do', 'what they would have wanted'. This coalescence of intersecting factors highlights the 'scapes' dimension of consolation explored in this chapter (and volume), consolation as a multidimensional arena of contact and exchange, whereby varied relational factors intersect, are expressed, mapped on to, and experienced in and through particular places, including bodies, communities, landscapes, faithscapes and the internet. The next section outlines a conceptual framework for mapping grief and consolation; it is followed by analysis of consolatory narratives in the experience and meaning-making of kin agreeing to a family member's organ donation.

Mapping grief and consolation: a conceptual framework[1]

> [The challenge]: how to render intelligible the perceptible and imperceptible qualities of space and spatial relations shaped by bereavement, grief and mourning; how to provide a framework that brings to light the spatial relations which underlie emotional geographies of grief, mourning[,] remembrance [and consolation] without objectifying them; how to reveal the interrelation of the material and emotional-affective, cognitive and the sensory, the individual or group and their wider social-cultural contexts[?]
>
> (Maddrell 2016: 169)

Bereavement is a near universal experience, but one which is experienced differently by individuals, families and communities, depending upon the assemblage of co-constituting intersecting factors such as relation to the deceased, age, gender and religious beliefs, if any. As I have argued elsewhere: '.... for some mourners and some deaths, grief is a mantle worn for a season and then shed in due course, for other mourners and other deaths, grief is both inhabited and inhabiting, strands of which can be woven into one's very being, forever changing emotional and affective DNA, shaping and influencing experience of the world. This is not to essentialize grief or the bereaved as grief-stricken and incapacitated, not least as grief can be an inspiration and catalyst, but rather to recognise the intertwining of loss in one's ever-emerging self and relations with others, as well as places and practices' (Maddrell 2016: 172).

Places that have or take on meaning in relation to the dead can therefore act as a catalyst, evoking memories, loneliness and/or comfort – or an unpredictable combination thereof. Moreover, the primary space of mourning is embodied by the mourner themselves, they carry grief *within* and can potentially be interpellated by it at any juncture of time-space. Grief, mourning and consolation are thus inherently spatial, i.e. experienced within and

1 This section draws in particular on my open access paper Maddrell A., (2016) Mapping Grief: a conceptual framework for understanding the spatialities of bereavement, mourning and remembrance, *Social and Cultural Geography* 17(2): 166–188.

though space, including the space of the body, and can be both *triggered* and *ameliorated* through comfort found in relation to specific places at particular times, producing internal emotional-affective 'maps' (Maddrell 2009a, 2009b, 2011, 2012, 2013, 2016). These individual and collective maps of grief and consolation 'map the territory' of a given bereavement (or other form of loss), and serve as navigational aids and route maps for those living with loss. Understanding the spatial dimensions of relational and overlapping grief and consolation offers insight to experiences of loss and ways of living with loss. Death and bereavement produce new and shifting emotional-affective geographies, whereby artefacts, places and communities can take on new and heightened significance. They can produce a whole new set of emotional topographies, mobilities and moorings, which can variously engender grief and consolation (Maddrell 2016). Understanding the complex dynamic internal emotional maps shaped by good and bad memories, lost futures and the unfolding experience of grief will help the understanding of individual and collective experience of bereavement.

The conceptual 'map' in Figure 2.1 is a schematic representation of the hard-to-capture overlapping and shifting assemblage of self-body-place-society that constitutes culturally inflected individual and shared spaces of grief and/or consolation. The 'Mapping Grief and Consolation' framework highlights the ways in which grief, mourning and remembrance are experienced in and mapped upon (i) physical material spaces, including the public and private arenas and artefacts of everyday life; (ii) the embodied-psychological spaces of the interdependent and co-producing body-mind; and (iii) the virtual spaces of digital technology; religious-spiritual beliefs; and non-place-based community.

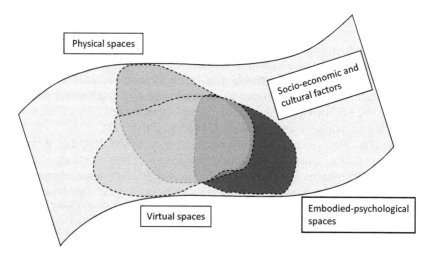

Figure 2.1 Mapping Grief: a conceptual framework for understanding the spatialities of death, dying, bereavement, remembrance *and consolation* (source: after Maddrell 2016: 181)

These maps can provide insight to the geographies of emotionally 'safe' and 'unsafe' places at a given juncture (Maddrell 2016: 166), including places where the bereaved find comfort and consolation and therefore constitute therapeutic environments (see Bell et al. 2018). However, the affective realm is pre-discursive, so while sites of emotional import may be predicted and 'mapped' in advance, affective responses may be unanticipated and sites associated with these only identified in retrospect.

Crucially, particular spaces and places can engender consolation as well as mourning. At any point in time particular locations may be sites of consolation, and these may or may not coincide with places engendering a sense of loss (see Figure 2.2). Wearing a deceased spouse's pullover may be simultaneously a material embodiment of loss and source of comfort; tending a grave or visiting the park associated with the deceased can engender the bittersweet emotions of the simultaneous absence-presence of the deceased; a bereavement support group may offer consolation through informal storying of the deceased (Maddrell 2016). As with grief, such mappings of consolation are dynamic.

Maps of grief-consolation should not be seen as static but rather as processual (Maddrell 2016): while particular places and events may be consistent in bringing emotional pain and/or comfort, others may advance and recede in their significance at different times. Such maps are shaped over time by intersecting factors including anniversaries, subsequent bereavements, new relationships and experiences throughout the life course. Despite Twentieth Century Western discourses which represented post-bereavement 'closure' as normative, the negotiation of living with loss can be a life-long engagement, likewise the related processes and practices of consolation.

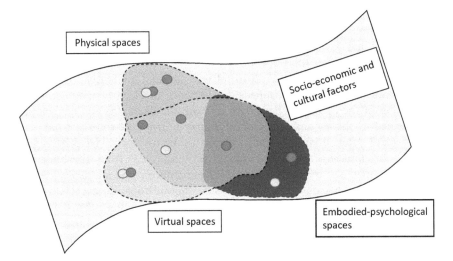

Figure 2.2 Overlapping spaces of grief and consolation at a given time

Furthermore, the cultural inflection highlighted in the figures above is to be stressed: grief, mourning, remembrance and consolation are shaped by wider social-cultural relations such as ethnicity, gender and religion (see Field et al. 1997; Kong 1999; Bhugra and Becker 2005; Maddrell 2011, 2016). The complex emotional geographies of lived place-temporalities, are coloured and shaped by socio-economic, cultural and political norms. This has implications for the character of consolation and the repertoire of consolatory spaces and practices available to particular mourners. For example, cultural expectations may *require* participation in or the funding of certain ritual practices or forms of hospitality (see Dunn et al. 2016 on Tanzanian rituals). Likewise, a refugee *cannot* return 'home' to attend a funeral; a parent can't visit an adult child's place of death on a battlefield; gender may exclude one from certain traditional social or religious rituals. In such circumstances other sites and practices of consolation must be found or devised.

The following section explores the discourses of grief-consolation in relation to deceased organ donation, a process which is typically associated with premature death.

The bittersweet consolation of organ donation: #livesoninothers

> Our little girl died ... but her kidneys saved another life.
>
> (Manger 2018, *The Mirror*)

In the context of ageing populations and declining donation rates in Europe, there is a shortfall between the availability of organs and the number of patients needing transplants; this shortfall is further extrapolated across the world, creating a bio-political divide between those who can and those who can't access transplants (Davies 2006). Numerous states are addressing this shortfall through governmental initiatives such as the promotion of deceased organ donation, payments to families for donated organs, and proposals to change national laws to an *assumption of agreement* to donation which requires opt-out rather than opt-in.

A recent review (Ralph et al. 2014) of international qualitative research on the nationally and culturally inflected experience of elective organ donation by families of the deceased sheds light on both the dilemmas and consolation experienced by those considering organ donation from their kin at what is an acute – and often unanticipated – period of bereavement. Healthy organs suitable for donation are often sourced or 'harvested' from the bodies of younger people, e.g. those who have died prematurely and suddenly, such as through a motor accident, making decisions about donation especially challenging for kin in the face of sudden bereavement. For those thrust into the shock of sudden bereavement and the simultaneous (often time-sensitive) need to make decisions about donating the organs of their deceased child, partner or sibling, this intersection of bereavement and organ donation is momentous and life-changing to the core; for some it is impossible to process in the often short time available. The following discussion focuses on the themes of comfort and consolation experienced through deceased organ donation, set within the context of discourses of corporeality and concerns about social attitudes, end of life care, the integrity of the body, and religious beliefs including responsibility to and respect for the body and spirit or soul of the deceased.

Finding 'meaning' through organ donation is at the centre of understanding people's willingness to allow organs to be harvested from the body of their kin. This meaning-making is broadly attributed to four core interpretations: i) saving lives; ii) letting the donor 'live on'; iii) fulfilling a moral obligation to society; and iv) easing grief (Ralph et al. 2014). Thus the links between deceased organ donation and consolation are clear. In a Greek study the bereaved reported experiencing organ donation as providing them with 'relief, tranquillity and a sense of purpose' through focusing on the positives derived from helping another to live, and as a distraction from their own loss (Bellali and Papadatou 2006; Ralph et al. 2014: 927, 928). This experience was echoed by a US relative: 'I think it gives me something more to think about besides death. This has diverted my thoughts to something positive' (Bartucci 1987, cited by Ralph et al. 2014: 928). Likewise, in the UK, families who opted for organ donation after the death of a child in a hospice found their decision helped them to cope with their loss. This view was echoed by parents in *The Mirror* newspaper headline quoted at the top of his section: 'Our little girl died ... but her kidneys saved another life'. For the family of an adult male, organ donation was described as bringing them comfort and even an emotional 'high' at a very bleak time (Carey and Forbes 2003; Ralph et al. 2014: 928). One bereaved family member in the UK threw a spotlight of self-criticism on their own motivations for agreeing to organ donation, stating: 'Its selfish really, because I wanted a bit of him to go on living you see' (Sanner 2007; Ralph et al. 2014). Like grief and consolation, altruism is complex, and it is widely acknowledged that, however motivated, altruistic acts may accrue benefits to the benefactor (see Barnett and Land 2007) – but this does not negate the generosity, nor its benefits. For some relatives, organ donation and its likely benefits to others was the *only* source of consolation in the face of loss, as reported by two sets of parents in a Canadian study (Long et al. 2006); this reflects the limitations to consolatory discourses and practices, especially in the immediate aftermath of tragedy, disaster or violence.

In numerous studies a sense of consolation was recurrently expressed by family members agreeing to donate their kin's organs, specifically in terms of a sense of the *continuity* of the life of the deceased. This highlights the significance of donation in the formation of what Davies (2002) describes as the social memory of the deceased, whereby the name and biography of the deceased is perpetuated. This was exemplified in a Canadian study which reported that all participants 'believed that organ donation was a means of somehow making sure this person's memory continued. The deceased relative's existence continues in some form, and in this sense helped keep the memory of the donor alive' (Jacoby et al. 2005; Ralph et al. 2014). This highlights the possibility of two aspects of 'lively' memorialisation through deceased organ donation: first, keeping the *memory* of the deceased alive within a wider circle which includes the recipient and their social network; and second, keeping a *physical part* of the deceased alive through the donated organs which, in helping others to live, continue to 'live' themselves. This moves beyond both cryogenic storage which preserves the dead body in stasis, and trees, shrubs and flowers which are planted as 'living memorials', sometimes rooted in and fertilised by the grave or cremated ashes of the deceased.

As one Greek mother's account testified, the corporeality and 'life' of the deceased continues through the person of the organ recipient, in this case literally in the form of the beating heart, the organ which defines physical life: 'It comforted me because although my child was buried, I was telling myself that he is still alive. What mainly helps me is to know that his heart is still beating' (Bellali and Papadatou 2006; Ralph et al. 2014: 928). Taiwanese Confucian and Buddhist families expressed a more explicit sense of relation with organ recipients, choosing to ascribe the organ recipients the status of members of their extended *family*, by dint of the biological connection through the organs and therefore life of the deceased (Shih et al. 2001, Tong et al. 2006). This sense of kinship network with the organ recipients engendered consolation for bereaved parents in a culture which privileges familial relations.

In some organ donation networks contact is encouraged between donor families and recipients, in the UK this is typically mediated through the National Health Service which can provide donor families with anonymised information on, and forward letters from, recipients. Interestingly, Ralph et al.'s (2014) overview of studies showed that Christian donors typically sought information about recipients' quality of life and health, highlighting how knowledge of the value derived by beneficiaries as a result of the donated organ offered an added sense of consolation. However, some donor families across national and cultural contexts preferred not to know about organ recipient outcomes for fear of disappointment if organs were rejected or the recipient died despite the donation, as this would undermine the consoling narrative of the deceased's contribution and liveliness.

The Organ Donation Isle of Man Memorial Garden

Returning to the example of the memorial garden with which this chapter opened, in this final section the public spaces, narratives and practices of a new organ donation network in the Isle of Man are discussed with reference to the mapping grief and consolation.

The space and narrative of the organ donation memorial garden at Noble's Hospital, Douglas, as well as the wider ethos and educational work which underpinned it, is grounded in generosity in extremis and the spaces and practices of consolation which that generosity engenders. The enclosed garden is one of several courtyards included in the hospital design, each providing light and green space for patients, staff and visitors, part of the therapeutic attributes of the building design. The garden was developed after a period of fundraising and is part of a wider initiative to promote organ donation in the Isle of Man. This initiative has been spear-headed by a mother whose teenage son died in a car accident and whose organs were donated to four other people. Diane described the garden and its purpose for families of the deceased, highlighting it as a space of recognition and pride: 'After 10 months of fund raising we were able to complete the memorial garden for donors from the Isle of Man. A special place to recognise and give renewed pride to families'. (https://www.linkedin.com/pulse/organ-donation-memorial-garden-nobles-hospital-isle-man-diane-taylor, accessed 6/2/18). Elsewhere, she described the garden as a 'fitting tribute' to the deceased donors. The Twitter feed for the organisation featured a photo of her son's memorial stone in the garden accompanied by the hash tags: '#GoneButNotForgotten #love #misshimsomuch #livesoninothers #myson'. This post echoes the emotions expressed in the international studies of deceased organ donation discussed above, as well as illustrating the intersecting material, embodied-psychological and virtual spaces of grief and consolation experience and practice also discussed above.

The memorial garden has two central stones, surrounded by five stones featuring laser etched images from around the Isle of Man (see Figure 2.3). Each donor has their own smaller named stone as a micro-memorial, with families choosing where to locate their loved one's memorial, for example by one of the place-focused etched stones allowing regional identification with an

Figure 2.3 Example of the Manx symbol (Loaghtan sheep) and micro-memorials in the Organ Donation Memorial Garden, Nobles Hospital, Isle of Man

area in the island, a Manx symbol (such as the Loaghtan sheep in Figure 2.3), with other known donors (kith or kin), or a favourite spot elsewhere in the garden. Since opening, the garden has been enhanced with additional ornaments, benches and plants, each materialising remembrance and representing the more-than-representational emotions of love and respect. A Sunset rose and potted perennials would bring colour in season, a birdbath attract birds and insects, giving a degree of liveliness to the enclosed courtyard; an armed services remembrance day poppy wreath linked one of those memorialised with wider military networks, and several vernacular carved items, including a large bench memorialising three donors, represented a local micro-network of kith and kin. The garden may be used for scattering or interring ashes but it is more typically an example of cenotaphisation, memorialisation without bodily remains (Kellaher and Worpole 2010). The garden thereby offers an additional space in which to name and story the deceased, as well as placing them within this honorific community of altruism. Likewise, it acts as a locus for a common community of experience for the bereaved who visit and tend the garden.

The memorial garden physically grounds the Organ Donation Isle of Man (IoM) initiative which uses print and social media to promote organ donation and build a network of supporters. Social media platforms such as Facebook and online memorial websites have become common sites of memorialisation, storying the deceased, sharing memories and expressions of continuing bonds (Kasket 2012; Maddrell 2012). The Organ Donation IoM Facebook, Twitter and LinkedIn pages provide spaces where information on and news relating to organ donation is posted to reach different audiences, and members can share their experiences. As noted above, 'storying' the deceased and narrating one's own experience of loss and carrying on constitutes a cathartic practice for many, witnessed by traditional obituaries and contemporary online memorials and bereavement blogs (see Maddrell 2012).

This storying is exemplified by the quote above (shared on the Facebook page of Organ Donation Isle of Man), whereby bereaved parents used a newspaper article to highlight the need for organ donation, reporting parents leading by example through donating their daughter's organs even though she herself died while on the waiting list for a donated organ (Manger, 2018, *The Mirror*). In the more local context of the Organ Donation IoM network, Diane wrote on the group's Facebook 'Story' section:

> In memory of my son [...] - Daniel donated his organs after we lost him aged 15. As his parents we made the decision to donate his organs x About me: I am just a normal (my friends would say otherwise :)) Mum with no medical back ground. I have set up this page to hopefully get people talking about this very sensitive subject. .... Only through the tragic loss of Daniel have I been able to hear about the success of organ donation. My page is not to say you must sign up its to discuss your wishes with your loved ones so they know what you would want. We are

a small Island and as you will see in one of my posts there are people waiting for transplants over here. Having read letters from the people who received Daniels organs their lives before transplant revolved round hospital visits and medication. I can't imagine how it was for them and their families. Thank you for taking the time to read this.

(https://www.facebook.com/pg/OrganDonationIsleOfMan/ accessed 6/2/18)

The post also invites questions and sharing experiences; and the cathartic benefits of storying the deceased can be seen in two other Organ Donation IoM Facebook page posts by mothers, both relating to correspondence from recipients via the NHS. One recounted: 'I received a thank you card this morning from a gentleman that has had his life turned around by a decision I made after losing my son. It was a difficult decision to make at the time, I won't lie, but the best decision I've ever made. Although I'd give anything to have my son in my arms, knowing there is a little of him still helping someone brings a little warmth to my broken heart'. She went on to plead: 'Please get registered! It takes two minutes and changes lives x' (ibid.).

Such posts encourage – give permission – to others to share their experiences, and the second mother's post gave a similar account to that above, highlighting her own dialectics of grief and consolation as well as the interwoven sympathy and gratitude from recipients and their family members. In this case the mother gave a moving testament to her affection through her words and emoticons, and gave tribute to her daughter, characterising her as a 'superhero' for saving someone else's life:

'So yesterday [I] came home from the workplace to find [a] big envelope from the Organ transplant team. Inside is 2 cards 1 from the liver transplant recipient and 1 from her daughter offering their deepest sympathy and gratitude for my Barb donating her liver which has given [Jane*] the recipient the gift of life and now her daughter [Ann*] can carry on life with her mama without worry[.] My Barb the superhero' (ibid., *pseudonyms).

It is interesting to note the role of *mothers* in this process of narration and sharing through the affordances of this online space, indicative perhaps of a gendered willingness and need to express themselves, and to memorialise their child in this public forum, a topic which merits further study.

Conclusion: Reflections on mapping grief and consolation

The relational, dynamic and culturally inflected emotional geographies of mapping grief and consolation are complex, dialogic and dialectic. The mapping grief and consolation framework outlined here offers a means of being attentive to the dynamic interplay of loss and consolation as variously experienced in and expressed through the spatialities of particular material places and artefacts, the nexus of body–mind and the virtual arenas of social media, online networks, and communities of experience and belief.

Under specific circumstances the opportunity to agree to deceased organ donation can afford bereaved kin with additional avenues within the broader repertoire of sites and practices of grief and consolation. Bereaved kin report consolation as drawn, to varying degrees (at different times and in different contexts), from a sense of moral obligation and helping others (including, for some, knowing something of the beneficiaries' outcomes); taking positive and life-affirming action in the face of an irreversible bereavement; and for some, a sense of 'living memorialisation' in the lives of the organ recipients, or even continued corporeal liveliness for the deceased centred on the partial continuity of their material body. This can be seen as a form of negotiating the relational absence-presence of the deceased donor, who is simultaneously dead, yet alive, giving an additional vector for continuing bonds through the psycho-emotional sense of something of a loved one still being 'in the world'. For some, this also represented an extension of kin networks through the materiality of the donated organs, and this brought comfort, especially in cultures incorporating ancestor veneration. The challenges to and potential benefits accrued from familial agreement to deceased organ donation highlights in sharp relief the bittersweet dialectics of grief and consolation, and their persistent and shifting temporal-spatialities. The spatialities of mapping grief and consolation across physical, embodied-psychological and virtual spaces were evidenced in the case study of the Organ Donation Isle of Man network. In this case various social media platforms and individual's posts, the physical garden and the embodied use of and care for the garden serve as a mesh of overlapping and intersecting emotional-affective spaces of commemoration and consolation. In turn these serve as a platform for the educational-political purpose of representing, promoting and lobbying for much needed organ donation in the wider community, linking personal loss and the possibilities of social change – and better lives for others.

References

Barnett, C. and Land, D. 2007. Geographies of generosity: beyond the 'moral turn'. *Geoforum* 38/6, 1065–1075.

Bartucci, M. R. and Seller, M. C. 1986. Donor family responses to kidney recipient letter of thanks. *Transplant Proceedings* 18, 401–405.

Bartucci, M. R. 1987. Organ donation: A study of the donor family perspective. *Journal of Neuroscience Nursing* 19, 305–309.

Bell, S., Foley, R., Hougton, F., Maddrell, A., and Williams, A. (2018) From therapeutic landscapes to healthy spaces, places and practices: A scoping review, *Social Science & Medicine* 196, 123–130.

Bellali, T. and Papadatou, D. 2006. Parental grief following the brain death of a child: Does consent or refusal to organ donation affect their grief? *Death Studies* 30, 883–917.

Bhugra, D. and Becker, M. A. 2005. Migration, cultural bereavement and cultural identity. *World Psychiatry* 4/1, 18–24.

Carey, I. and Forbes, K. 2003. The experiences of donor families in the hospice. *Palliative Medicine* 17, 241–247.

Davies, D. J. 2002. *Death, Ritual and Belief.* London: Continuum.

Davies, G. 2006. Mapping deliberation: calculation, articulation and intervention in the politics of organ transplantation. *Economy and Society* 35, 232–258.

Dunn, C.E., Le Mare, A. and Makungu, C. 2016. Connecting global health interventions and lived experiences: suspending 'normality' at funerals in rural Tanzania. *Social & Cultural Geography* 17/2, 262–281.

Field, D., Hockey, J., and Small, N. 1997. Making sense of difference: Death, gender and ethnicity in modern Britain, in *Death, Gender and Ethnicity*, edited by D. Field, J. Hockey and N. Small. London: Routledge, 1–28.

Jacoby, L.H., Breitkopf, C.R. and Pease, E.A. 2005. A qualitative examination of the needs of families faced with the option of organ donation. *Dimensions of Critical Care Nursing* 24, 183–189.

Kasket, E. 2012. Continuing bonds in the age of social-networking: Facebook as a modern-day medium. *Bereavement Care* 31, 62–69.

Kellaher, L. and Worpole, K. 2010. Bringing the dead back home: urban public spaces as sties for new patterns of mourning and memorialisation, in *Deathscapes: New spaces for death, dying and bereavement*, edited by A. Maddrell and J. Sidaway. Farnham: Ashgate, 161–180.

Klass, D., Silverman, P. R. and Nickman, S. L., eds. 1996. *Continuing Bonds: New Understandings of Grief.* London: Routledge.

Klass, D. 2014. Grief, consolation, and religions: a conceptual framework, *Omega* 69/1, 1–18.

Kong, L. 1999. Cemeteries and columbaria, memorials and mausoleums: narrative and interpretation in the study of deathscapes in geography, *Australian Geographical Studies* 37, 1–10.

Long, T., Sque, M. and Payne, S. 2006. Information sharing: its impact on donor and nondonor families' experiences in the hospital. *Prog Transplant* 16, 144–149.

Maddrell, A. 2009a. Mapping changing shades of grief and consolation in the historic landscape of St. Patrick's Isle, Isle of Man, in *Emotion, Culture and Place*, edited by M. Smith, J. Davidson, L. Cameron and L. Bondi. Ashgate: Aldershot, 35–55.

Maddrell, A. 2009b. A place for grief and belief: the Witness Cairn at the Isle of Whithorn, Galloway, Scotland. *Social and Cultural Geography* 10, 675–693.

Maddrell, A. 2011. Bereavement, belief and sense-making in the contemporary British landscape: three case studies, in *Emerging Geographies of Belief*, edited by C. Brace, D. Bailey, S. Carter, D. Harvey and N. Thomas. Cambridge: Cambridge Scholars, 216–238.

Maddrell, A. 2012. Online memorials: the virtual as the new vernacular. *Bereavement Care* 31/2, 46–54.

Maddrell, A. and Sidaway, J. 2010. Introduction: Bringing a spatial lens to death, dying, mourning and remembrance, in *Deathscapes: New Spaces for Death, Dying and Bereavement*, edited by A. Maddrell and J. Sidaway. Farnham: Ashgate, 1–16.

Maddrell, A. 2013. Living with the deceased: absence, presence and absence- presence. *Cultural Geographies* 20/4, 501–522.

Maddrell, A. 2016. Mapping grief: a conceptual framework for understanding the spatialities of bereavement, mourning and remembrance. *Social and Cultural Geography* 17/2, 166–188.

Manger, W. (14/2/2018) Our little girl died waiting for a heart but her kidneys saved another life. *The Mirror*, https://www.mirror.co.uk/news/uk-news/little-girl-died-waiting-heart-12029096 (accessed 6/3/18).

Martin, T. L. and Doka, K. J. 2000. *Men Don't Cry ... Women Do: Transcending Gender Stereotypes of Grief*. London: Brunner/Mazel.

Ralph, A., Chapman, J. R., Gillis, J., Craig, J. C., Butow, P., Howard, K., Irving, M., Sutanto, B. and Tong, A. 2014. Family perspectives on deceased organ donation: thematic synthesis of qualitative studies. *American Journal of Transplantation* 14, 923–935.

Robinson, J.D. 2010. Leathean landscpaes: forgetting in late modern commemorative spaces, in *Memory, Mourning, Landscape*, Edited by E. Anderson, A. Maddrell, K. McLoughlin and A. Vincent. Amsterdam: Rodopi, 79–97.

Sanner, M. A. 2007. Two perspectives on organ donation: experiences of potential donor families and intensive care physicians of the same event. *Journal of Critical Care* 22, 296–304.

Santino, J. 2006. *Spontaneous Shrines and the Public Memorialization of Death*. New York: Palgrave Macmillan.

Shih, F. J., Lai, M. K., Lin, M. H., *et al.*2001. Impact of cadaveric organ donation on Taiwanese donor families during the first 6 months after donation. *Psychosomatic Medicine* 63, 69–78.

Siminoff, L., Mercer, M. B., Graham, G. and Burant, C. 2007. The reasons families donate organs for transplantation: Implications for policy and practice. *Journal of Trauma* 62, 969–978.

Stevenson, O., Kenten, C. and Maddrell, A. 2016. Editorial: And now the end is near: enlivening and politicising the geographies of dying, death and mourning. *Social and Cultural Geography* 17/2, 153–165.

Stroebe, M. S. and Schut, H. 1999. The dual process model of coping with bereavement: rationale and description. *Death Studies* 23, 197–224.

Tong, Y.-F., Holroyd, E. A. and Cheng, B. 2006. Needs and experiences of Hong Kong Chinese cadaveric organ donor families. *Journal of Nephrology* 8, 24–32.

Valentine, C. 2008. *Bereavement Narratives: Continuing Bonds in the Twenty First Century*. London: Routledge.

Walter, T. 1996. A new model of grief: bereavement and biography. *Mortality* 1/1, 7–25.

https://www.linkedin.com/pulse/organ-donation-memorial-garden-nobles-hospital-isle-man-diane-taylor (accessed 6/2/18).

Part II
European constellations

3 Consolation and the 'poetics' of the soil in 'natural burial' sites

Albertina Nugteren

Introduction

The spirit behind 'green' death practices has caught on in many areas world-wide. The catalysts for such practices may be threefold. First, the environmental consequences of conventional burial become untenable to those who are environmentally conscious and whose 'green' lifestyles prompt them to opt for 'green' deathstyles. Widespread use of toxic fluids conserving the body in concrete-lined pits may be objected to in the US; lacquered plywood caskets full of chemical glues or the highly toxic state of today's deceased are targeted in other parts of the world. But these are just a few of the practices that are resisted (Basmajian and Coutts 2010; Canning and Szmigin 2010; Feagan 2007; Harris 2007; Nugteren 2015b; Schade 2011). There are obviously many more.

The second motivation may be less activistic and tangible but is certainly as compelling. There are those who are driven by a spiritual and aesthetic need to reconnect themselves to nature while alive; they likewise want to get 'back to nature' in death (Kaufman 1999; Rumble 2010; Davies and Rumble 2012; Clayden 2013; Nugteren 2014a; 2015a and b). In a preference for a 'dust-to-dust' or 'earth-to-earth' attitude they resist being 'insulated' from natural cycles of decay, transition and rebirth. Instead, the idea of their bodies gradually decomposing and thus nurturing new life forms gives them a deep sense of solace (Feagan 2007; Klaassens 2011; Taylor 2010: 13).

Third, and more fundamentally, the 'greening' of disposal practices may be seen as a rejection of culturally ingrained divides between body and soul, or between humans and nature, and, as we shall see later on, between the soul, the self and the soil.

In a wider historical and geographical perspective there is nothing really new about natural burial. In fact, as far as we know, this is more or less how it has always been done. When people speak of natural burial as 'new' practices today, this is true merely insofar as these are a reaction to a death industry in the Western hemisphere that is hardly more than a century old (Ariès 1981; Davies 2002 and 2005; Harris 2007; Venbrux et al. 2013). On a deeper and more fundamental level, the 'greening' of disposal practices may be seen as a subcultural or even countercultural rejection of much older divides between a body which had, in a culture-specific way, long been considered temporary (and thus thoroughly problematic) and a (winged, disembodied) eternal soul. Whereas contemporary death culture in the Western world may be far less explicit about post-death survival of a culture-specifically defined and imagined soul, today a kind of figurative or symbolic immortality appears to have taken over (Lifton 1979; Kearl 2010; Klass 2006). The actual funeral and cremation experience may often be 'merely' an occasion for revisiting and reworking memories of the deceased's life-story (van Tongeren 2004; Venbrux et al. 2009). However, grave-sites and columbaria are not the only public spaces in which the personal and private dimensions of grief are sentimentalised. The same is true of newspaper announcements, obituaries, digital condolence sites, virtual memorials, park benches formally inscribed by donors in memory of a loved one, and roadside memorials to victims of traffic accidents. Sites and signs of private memorialisation are increasingly part of public space. Tributes to the lives of public figures alternate with idiosyncratic tributes to individuals who apparently had a special affinity with a particular location. As passers-by we read their celebratory life-stories on plaques 'storying the deceased' (Walter 1996), or, conversely, wonder about the cryptic references to someone never heard of before. Public mapping of private emotions thus creates 'memoryscapes' and 'threshold spaces' distanced from home and cemetery or crematorium. Immortality, in secular society, thus means: being celebrated, remembered and memorialised not only by one's significant others but also by any casual passer-by. For a moment, casual neutral space turns into a transformative space where the borders between one's own vigorous life – running, trekking, walking, strolling – and someone's death indicated by a memorial bench become temporarily porous.

In this process, somehow the real-life physical image of the deceased gets either frozen in time (i.e. a few months or even years before death), or retraced to what might be considered the beloved one's apex, the highest point of an individual life's arch, the person at his or her 'essential' best. The story of what actually happens to the interred body is hardly ever told, imagined, or otherwise referred to. The commemorative image we carry of the deceased is largely static. This shying away from the dead body's abject realities may indicate one of the most persisting taboos or sensitivities of contemporary Western culture. In spite of various paradigmatic shifts in which the human body and its functions became objects of scientific investigation, surgical engineering, chemical manipulation and neuro-processual analysis, all with their concomitant disenchantments, a culture-specific no-go area has re-emerged where our imaginings are not allowed to linger: the physical 'indignity' of the decomposing body. Instead, there appears to be a consensus about the piety we owe the deceased, any deceased, even an anonymous one. The dignity of the dead should be protected by all means, and particularly from our all-too realistic thoughts.

Some, however, voice alternative views and perform alternative practices. What is considered 'mainstream' and what 'alternative' may be subject to considerable shifts even within a single generation. The fact that cremation, for instance, by now has become mainstream should not obscure its troubled origins. The vigorous opposition it once encountered from various quarters in Western countries (Bucknill 1915) is almost unimaginable a century later. One of the internal paradoxes of European culture may be found in the fact that it prides itself on its reasoning or rationality while people everywhere blissfully un-reason themselves in a few specified – culturally sanctioned – fields: love, art, mysticism, death. In much the same way, the human body finds itself in a precarious balance between irreversible disenchantment and counterforces of re-enchantment popping up, sometimes from unexpected directions (Asprem 2013; Gibson 2009). This multidirectional interplay of cultural, subcultural and countercultural vectors may also be perceived in the Green Death Movement.

In this chapter I take those cultural and spatial considerations into account in order to explore what layers of meaning and significance may be attributed to the term 'nature' (or its derivative, 'natural'), as in the phrase 'natural burial sites'. In focusing on such sites, I detect various shades of green, covering the entire range from deep-ecology, deep-time considerations to a trendy 'naturalness', and from nature as 'something out there' to nature as 'all there is', emphatically including our human existence. In addition, I address the mechanics of consolation. Both at the collectively cultural level and at the individual level, rifts between concepts of consolation divide contemporary practices of death and disposal. What may be consoling to most – the vision of a 'positive mass' of family and friends, neighbours and colleagues, late-in-life casual acquaintances and an empathetic ritual facilitator who is delicately able to include them all in a service to celebrate a life well-lived – may be a disheartening prospect for others. The feelings of comfort that may be found – both by the person imagining such a commemorative post-life gathering for him- or herself, and the persons who assemble there to grieve, pay tribute to the deceased, and derive solace from shared reminiscences – may be paralleled in the experiences surrounding the solemn disposal in a graveyard (Hockey et al. 2010). Precisely at moments of shock and loss, grieving and rupture, the joint action of putting the deceased 'to rest' in a decent graveyard, with neatly trimmed hedges, an occasional evergreen, and well-defined footpaths preferably accessible to wheelchairs and rollators, is a well-choreographed middle part of the emotional 'curve' between the commemorative service and the final refreshments. Why would one want it differently?

Environmental concerns

The reactions to reports about the globe's rapid deforestation, pollution of earth, water and air, depletion of mineral resources including fossil fuels, anthropogenic climate change or shrinking biodiversity, vary considerably. To some people, the reports may seem alarming enough, but only in the long run. To others, the problem simply seems too massive for a single individual to become involved at all. But some feel personally challenged and are prompted to behave more responsibly themselves and to urge those around them to do likewise (Fromm 2009). Awareness-raising campaigns about the intricate system of which we humans are a part, the biosphere in which we live through inter-being, and the democracy of all beings that we should tune in with, may address a religion-specific sense of responsibility. Alternatively, they may advertently steer away from any appeal to pre-existing worldviews and remain 'strictly sci-entific'. Some organised religions may indeed be 'greening': by selectively pointing at scriptural passages or inherited practices now interpreted as 'green', opinion-leaders set out to mobilise the faithful into more environmentally responsible behaviour. Such attempts may be successful to the extent that people mine the tradition they subscribe to in a search for inspiration in order to tackle a pressing contemporary problem. By remaining within their own heritage, and uncovering long-lost ecological wisdom in it, they perceive themselves as remaining true to their own master narrative. But since the grand narratives of established traditions are necessarily full of inner contradictions, inconsistencies and incongruities, the one-on-one application of a mythical scenario to con-temporary issues (Sheldrake 1991) may require considerable acrobatic skills, ranging from naively ignoring the anachronistic pitfalls of such an exercise to creative contortions on the tightrope between science and mythography. What-ever label we use for today's multiple modernities (Eisenstadt 1999), we still – in fact, all over again – live by our mythologies (Barthes 1972).

But on the individual level, at least, small things *can* be changed: patterns of consumption, patterns of behaviour, patterns of interaction with the non-human world. The idea that even their final footprint will be a burden to the environ-ment has created growing uneasiness among some segments of society. Even in death we contribute to pollution, we occupy precious space, we exercise our claim to dignity and remembrance by increasingly irresponsible habits: toxic chemicals, non-biodegradable materials, hothouse flowers, tomb-stones quarried in distant countries dodging even the most basic human rights. That such cul-ture-specific claims to dignity in death have a downside becomes increasingly clear (Albery et al. 1997; West 2010; Wienrich and Speyer 2003). Environmental concern about the Western death industry takes both a short-term and a longer perspective: available space, turnover cycles, hygiene, religion-specific require-ments, legal procedures about perpetual grave rights, etc. All these 'grave mat-ters' (Harris 2007) are added to doom scenarios about long-term consequences for the limits of natural resources and the vast cycle of regeneration needed for the deep imprints that contemporary death habits make on the planetary future.

The consolation that may be found in choosing a manner of disposal leaving less, little or no imprint on the natural habitat and making merely minimal use of this earthly planet's resources is obvious: it frees the conscience, to a degree.

A call 'back to nature'

Thinking less in negative environmental consequences of today's 'death indus-try', others opt for natural burial as a matter of taste, style, and aesthetics. Vaguely inspired by a romantic 'back to nature' trend they design their own disposal and remembrance in a lovingly landscaped setting: well-tended trees, waving grasses, the sparkle of a stream, clean air, bird-song (Rumble 2010; Schade 2011). Nature, thus, is a commodity (Castree 2003). It provides the preferred scenario for imagining one's own 'remains' after death, as well as for imagining how those left behind might find solace and significance in the natural beauty surrounding one's grave (Clayden and Dixon 2007; Davies 1988; Davies 2002; Davies and Rumble 2012). The depressing 'feel' of conventional, crowded, sanitised and manicured burial sites may be counteracted by deliberately choosing spacious sites of natural beauty. This may be realised by pre-registering in 'normal' cemeteries with available slots beneath mature trees. An alternative may be found in some of the scenic sites designated as natural burial sites (NBS) – often private property such as estate parks, orchards and farmlands, or protected 'national' landscapes. Many such places are now officially acknowledged and even certified as 'natural', 'green' or 'eco' burial sites. Another option is provided by terrains of so-called 'new nature': corners of land retrieved from former industrial use, demolished social housing projects, or abandoned farms. Such sites may still look desolate and bleak, but burials may gradually – hopefully – shape the site's outlook and over time co-determine the biosphere by grasses sown, indigenous plants and wildflowers brought back in, trees planted in remembrance and rocks heaped on graves.

In some cases the distinction between the three catalysts – environmental concern, aesthetic preference and a recovered dramatic sense of embodied 'earthiness' – is difficult to determine. My own empirical research shows that most 'death aesthetics' centre primarily on human needs, even when a natural setting is preferred over the manicured ambiance of conventional cemeteries. In other words, despite rhetoric about nature, human beings themselves still stand centre stage (Clayden and Dixon 2007; Groote and Klaassens 2010). From a more radical perspective this could be denounced as merely 'shallow ecology'. The centrality of human needs, human preferences, and human symbolism marks this aesthetically determined type of natural burial as a lighter shade of green[1] than the deep environmental concern about one's final footprint (Taylor 2010). Well-regulated beauty brings solace, and natural beauty may work on grief in various ways. It may help, in anticipation, to overcome the fear of death and the 'cold', 'chill' or 'dark' some civilisations associate with death. It may assist in focusing the attention on the life-forms around the grave instead of on

---

1 In a previous article, 'A Darker Shade of Green? An Inquiry into Growing Preferences for Natural Burial' (Nugteren 2015b), written with a North-American readership in mind, I distinguish three shades of green, i.e., three types of natural burial sites: (1) burial grounds rated as hybrid or mixed ('ranked' with 'one leaf', in a system resembling the 'stars' accorded to restaurants and hotels); (2) those sites generally termed natural burial sites (NBS), such as patches of mature woodland, but also former meadows, converted orchards, and so-called 'new nature' such as abandoned industrial locations or other dumps and wastelands, deserving a 'two-leaves' label; and (3) conservation burial grounds in national parks and stretches of wilderness, deserving 'three leaves'.

the interred body. The rhythm of the seasons as manifested in the cycle the surrounding landscape goes through may divert the attention from rebellious, lonely, sad, or even morbid thoughts by those who visit the grave. Such aesthetically pleasing burial sites, with nature seen in soft focus, may function as therapeutic landscapes (Gesler 1992; Williams 1999). In stages, these sites may help the grieving to cope with loss and transform acute pain into new symbols and significances (Attig 1996; Janowski and Ingold 2012; Kearns and Gesler 1998; Klass 2006; Schama 1995: 15).

Both grief and consolation practices may easily border on platitudes or even kitsch. The expressions of hope and remembrance found on conventional, often religiously inspired gravestones only differ from the above in the type of symbols they use and the type of master narratives they link up with (Kramer 1988). The promise of an eternal soul, physical resurrection at the end of time, or reincarnation in one form or another, may be referred to in terms of religiously specific images. Or, increasingly so, in broadly humanistic values of each individual's right to be remembered (Kunkel and Dennis 2003) and thus made immortal in a symbolic sense (Lifton 1979). On many accounts, the coping and consolation practices as observed at natural burial sites resemble both religious and humanist mechanisms. It appears that the urge to console may be as instinctive as the need to be consoled, even if none of the words and images are taken literally (Walter 1996). Many of those, even, become an expected drone, a background tune that merely functions as wallpaper to cushion raw grief and social awkwardness in the face of death. But at least there is one important distinction: neither religious constructions of hope nor humanist appeals to the fundamental right of human dignity define consolation in terms of life itself.

Bio-centric consolation, on the other hand, goes beyond both gods and humanity: it consoles by weaving the dead into the ongoing texture of all life-and-death that layers our planet (Feagan 2007). No heavenly frills or memorial embellishments are needed: the scientific certainty that one invests in this bio-scopic narrative is good enough in itself. We will now look more closely at this third perspective.

The biotic non-self

Peter Wohlleben, a German forester, suddenly found himself a celebrity when his book *The Hidden Life of Trees* (2017, German original 2015) became an international bestseller. He shares his deep love of woods and forests in a language that is both scientific and tenderly, humorously, empathically attentive to detail. Amazingly slow but smart processes of life, death and regeneration unfold before the reader's eye. Most of it may be common knowledge, but with Wohlleben's gaze it turns into a wonderland, especially when he gradually builds his case that the forest is an intricate social network. This, too, has been indicated by others. But where he leaves his reader truly gasping is when he dwells on the invisible: the forest soil. In a chapter (15) aptly named 'In the realm of darkness' he refers to the incongruence that we know less about the ocean floor than about the surface of the moon, and that we know even less about the complex life forms that bustle beneath our feet. It may be a hidden world – most people even abhor it – but up to half the biomass of a forest is there, especially in the form of the mycelial web. And then he hits the reader with a simple sentence (2017: 86): 'There are more life forms in a handful of forest soil than there are people on this planet.'

Why, then, do most people shy away from all-too-realistic views on the deceased's body left in that same wondrous soil? Or, as Wohlleben writes: 'To become part of the ancient forest after death – isn't that a wonderful idea?' (2017: 92). Why the need of culturally specific embellishments, euphemisms, taboos? Why the need of metaphors for the state of death in terms of winged souls, angels, butterflies, birds, rainbows, paradises beyond the clouds?

The human mind may make its loops and lassoes around reality, it may construct fantastic edifices of meaning, yet it always finds its counterpoint in the soil. And it appears that this soil has the final say when it silences the mind and takes dead matter as composting material for the ongoing life story, just as it has done for millions of years. Nature, as a cultural construct, is a representation, a symbolic form of a final and all-encompassing category, but soil is real, tangible, material. In its fragmented and framed form we call it 'nature', in lower case, but the word landscape – or better still: earthscape, soilscape – might be more fitting. Its fascinating history is written in paleontologically unearthed layers covering our planet, opening up to us petrified sediments, geological patterns, rocks and boulders fascinatingly inscribed by deep time, although plain soil is generally considered boring. And yet it is this very soil to which we have to entrust our deceased, and it is this same soil for which we ourselves are destined when our time has come. Most, at best, are only gradually able to write themselves or their deceased into that ongoing narrative of the earth's skin. If they bravely do so, consolation may be found in 'scoping' this specific individual's death by linking it to the greater story, the earth story, the soil story. Consolation may thus be found in a biophilic surrender according to which humans owe the very fact that they live to the death of previous life forms. The air they breathe, the earth beneath their feet, their food, their brilliant tools, as much as the fossil fuels in their vehicles: all exist because other entities existed before them. In front of an open grave, hoping and coping in a naturalistic way is thus a matter of scope. Scope, by a grave, means some form of acknowledgement that in layer after layer of decomposing life, the great story of matter into which all existences are woven, is their own story as well. The so-called self may be effaced and obliterated, but it becomes written into the soil. The soil story may thus well be the only story that holds.

In a strikingly candid and visionary way, this was articulated by the Canadian writer Robert Feagan in his essay 'Death to Life: Towards My Green Burial', when he imagines his preferred way of dying and decomposing by 'exposure':

> With my life force spent and my body slumped against a gnarled pine tree on the pre-Cambrian shield near the lake, carrion tear at, and flies lay their eggs in, my decaying flesh, while my fluids slowly drain into the

cracks in the ancient rock to enter the roots of that same tree against which I exhaled my last breath.

(Feagan 2007: 157)

In the author's earth-centred burial musings he speaks of 'narratives of the human-earth life-cycle' and green burial as 'evocative of a more comprehensive and spiritual ethos of connection, continuity, and responsibility' (Feagan 2007: 157). He makes a plea for a more thorough attention to the corporeal component of the transition that culture-specific phrases refer to (Feagan 2007: 159). Burial practices should be more explicit in recognising reciprocal responsibility and the continuity of life after death in terms of the basic functions of the biosphere. In the United Kingdom, Hannah Rumble likewise heard many musings and motivations about 'giving oneself back to nature' (Rumble 2010), both because it is a soothing image and it feels good for the sake of ecospheric integrity (Fox 2006). Some do so in terms of intergenerational equity – thinking and acting across generations (Feagan 2007: 163) – whereas others consider much shorter time-horizons. Narrow anthropocentric considerations of the environment ('shallow ecology', Naes 1973; Fox 1990) often disregard such integral connections – and obligations – to the biosphere, or to what Aldo Leopold famously called the 'biotic community' (Leopold 1970: 262).

However eulogised, most natural burial sites are merely 'managed landscapes', tamed and framed, 'molded into entities to do our bidding' (Willers 1999: 3). Their so-called nature is co-produced nature-culture, relational, processual (Giannachi and Stewart 2005: 19; Soulé 1995; Szerszynski et al. 2003: 4). The consolation that humans find in natural beauty, the solace that is derived from surrendering to something greater beyond them: both are processes that take the sting out of the defeat, the ugliness, the utter desolation. The person pre-registering for a self-selected spot may make peace with his/her own mortality by gradually starting to love the place. The persons left behind may find solace in the continuing life around the grave, life that springs up spontaneously or is induced by ritualising grievers who plant and sow, sprinkle and fertilise. In most natural burial sites grave markers and grave cultivation must be kept to a minimum whereas at other sites non-natural, non-biodegradable or non-indigenous memorial culture will be condoned during at least the first phase of grief (Nugteren 2014a). From the perspective of some mourners such a regime may be too strict: ritualising around the grave is a deeply felt human urge to bridge the awkward gap and to continue bonds. Other mourners respectfully let nature – small nature in the form of micro-organisms, seeds and shoots, worms and beetles, bird droppings, mushrooms, acorns and autumn leaves – do its job.

Scoping, coping and consolation

Various coping and consolation styles can thus be discerned. What means utter desolation to one – an untended perhaps even anonymous grave, overgrown with brambles or weeds, and hard to locate since 'proper' pathways are lacking – may be tenderly consoling to another. Coping, by a grave, in my view, appears to be 'scoping': the scope with which one can gradually, processually, relate to ongoing reality, and particularly to ongoing nature. From small signs of life emerging around the grave hope and perspective may be derived, and for those inclined to a deeper-shade-of-green mentality there may even be beauty and consolation in surrendering to the social network of the soil – in fact, to the entire web of life.

Cultures have spoken copious words against death (Davies 2002, and 2005: 110). Most cultures and civilisations are conspicuous by their genius to create buffers with which to soften adversities, loss and defeat, make meaning, and give positive spins to death. Apart from being a ritual process, grieving is also a cognitive process, and this is where the symbolic order is of paramount importance. There is, however, a major distinction in coping and consolation styles that 'go against death', and those that 'go along with death'. The choice in favour of natural burial relatively often shows a style that goes along with death. We notice a preference for a naturalistic view of death, a dust-to-dust and earth-to-earth attitude that intentionally breaks free from religiously and culturally inherited euphemisms and avoidances. In that sense the rising phenomenon of natural burial is both a rejection of existing disposal and memorial practices on the one hand, and a congruity between lifestyle and deathstyle, on the other. If people call life an enigma, death is a predicament. According to Harari (2016) *Homo sapiens* rule the world because only they are able to weave collectively ingrained webs of meaning. As human meaning-makers, we are expected to repair industriously, frantically and stubbornly wherever and whenever the fabric of life is broken. Some religious traditions build dams against death, but other traditions go along with it. A basic attitude of going along with death may be grounded in a matter-of-fact, no-nonsense attitude, but may just as well be the outcome of a sophisticated argumentative approach or a poetic 'realism'. The ability to go along with the flow in life may result in the ability to go along with death when and where it presents itself. Culture, as a dam against death, may sit in the way of a natural flow and acceptance, but there is no denying that at the same time it is a vast repository of fascinating constructions meant to soothe raw edges and redirect attention. And, admittedly, there is more than a family resemblance between such culturally inherited imagery and the need for a good story, the earth-story, the soil-story, to link up with while standing in front of an open grave.

Some say that humans have lost touch with the story they take part in. Even more so, it is often stated that they have intentionally stepped out of it, and set themselves apart, categorically. Where that paradigm – humanity at the centre of things – discloses the disastrous consequences of the Anthropocene, irrepressibly another category is on the rise. The category Nature may currently be in such a phase, and one of its referents, landscape, may help us to understand more about what is at stake in the natural burial/green death movement. On the surface one could call it a trend – a hype, even. In calling it a hype we can resort to habitual cynicism: hypes come and go, they are as shortlived and meaningless as fashions. Of course, for economy's sake governments, municipalities and entrepreneurs should accommodate this growing demand for natural burial opportunities (de

Haas and de Vries 2013), but the question whether, possibly, this phenomenon lays bare some fundamental cultural critique is mostly disregarded. This is why I address the phenomenon of a rising demand for natural burials as more than just another manifestation of the individualisation and personalisation of disposal practices. The desire to 'return to nature' in death (Kaufman 1999) is, from my perspective, not merely a matter of aesthetics and vague romantic longings.

When probing during in-depth interviews I often touch upon underlying notions of lost connections. A sceptic may counter this by pointing out that such regrets come a little late in life, as if it is only when death is approaching that people begin to realise the alienation they had long considered normal. This implies that the process of dying and death is unduly burdened with long-postponed fundamental insights, comparable to the way an 'in-the-face-of-death' acknowledgement of sins may function within religions. But, just like sin, ontological alienation is no small matter. In this register, the separation of humanity from the earth and the soil, as well as the separation of the soul from the body and the material world, are both perceived as ultimately destructive and alienating. Because the grand story of Planet Earth, so appealing to palaeontologists, may put off average Westerners, Feagan introduces bio-regional consciousness and sense-of-place as more fittingly matching the scale and scope that most people live with (Ashworth and Graham 2005; Bender 2006; Davidson et al. 2005; Heise 2008; Maddrell and Sidaway 2010; Smith et al. 2009). On one's own particular patch of earth it may come more naturally to feel connection and responsibility, and when this would translate to 'earth-conscious' disposal practices people might more easily become embodied and embedded (Feagan 2007: 165). Part of humanity's disorientation, however, both in life and in death, may be caused by modern humans' urban mobility, uprootedness, placelessness – in short, supermodernity: there are many for whom no place feels special or at home (Augé 1995). On the other hand, global awareness may be enhanced by regularly seeing Earth from a distance, such as through an airplane window, deep below. Earth itself can feel like an endearing and precious home when seen as a blue-green magic marble floating in space, an image we have become familiar with ever since space programs and satellites provided us with such reversed perspectives (White 1987; Nolan 2013).

Summary and tentative conclusion

So what about this small rectangle where our own grave is envisaged, or where the body of a dear one is buried? Being burdened by moral responsibilities towards the biotic community, or having to think through biospheric layers and processes, may well turn out too demanding for the average mourner. On top of that, having to visualise the reality of a decomposing body may cause panic and even trauma. Realism in matters of death is considered to be only for the brave, or for disinterested funeral professionals. Yet the above could provide a clue to a rethinking of death in both its natural and cultural connotations. The unease and cultural anxiety about our ecological footprint even in death, as displayed in the Green Death Movement, indicates a countercultural tendency, a resistance to common current practices. My long-term observations (Rose 2007) in such sites, supported by in-depth interviews with all kinds of stakeholders (Van der Aa and Blommaert 2015), result in a mixed image:

(1) Those who have led a lifestyle to which a natural burial would be a fitting finale, benefit hugely from their conscious choice. The idea that one would be buried under that tree, under those waving grasses, on that slope, in that slanting sunlight, amidst tiny creatures crawling the earth or working their way through the soil, may be a comforting thought and a beautiful way to go. The common expression of 'being put to rest' there may evoke visions of peace and fulfilment. Some do indeed connect to the larger picture of transitive decay, as a return to the biotic community to which one 'gives something back'. However, they do acknowledge that their body isn't worth that much, or may even burden the environment, especially when it is filled with chemicals after medical treatment or when it contains artificial devices such as implantations. In spite of such relativising remarks, those who opt for a natural burial derive much comfort from their pre-registration, both for their own sake and for the sake of their descendants, if only because both the place and its upkeep (sometimes even perpetual grave rights) have been prepaid.

(2) Some relatives feel that a grave in a natural burial site fits the individual styles and tastes and convictions of the deceased, so that the place becomes an apt expression of his or her life-story and personality. Visiting the place may feel like visiting the person, and the ongoing life around the grave may be experienced as pleasing to the deceased and as a significant symbol to the mourners. Memorial culture may be left to a minimum, as this may have been the express wish of the deceased or it may be because of the rules and regulations of the particular site. Some visitors, however, perceive natural graves as untended, neglected, desolate, especially in the beginning when the spot itself looks bare, bleak and ugly (Clayden et al. 2010; Walter and Gittings 2010). The urge to decorate and beautify, to care and tend, to plant and sow may be accommodated within certain limits, at least in the first phase of mourning. Some sites allow only local indigenous materials, others allow any biodegradable stuff. From prolonged fieldwork experience I learned that in such natural burial sites the first phase of mourning is most crucial and critical: the natural environment may either bring solace and perspective, indeed, or conversely enhance feelings of grief, especially in desolate weather. Without the usual buffer of neatness and tidiness, well-tended plants and flowers, a well-chosen headstone with the deceased's name and some endearing words on it, and the comfort of rows of evergreens walling off the cemetery from profane life, a natural burial site may feel shockingly unprotected. This unease may be enhanced by the awareness that the body was buried relatively shallowly, in a shroud or a biodegradable coffin, and is exposed to forces of decay in a way which may make the imaginative visitor highly uncomfortable.

(3) As soon as this first raw reality has been countered – or rather: as soon as the mourners have started to go along with death – a natural burial site can become a place of solace, comfort and consolation. Processes of ongoing life,

the rhythm of the seasons, the discrete sharing of the place with others of more or less the same values, the precious messages one may read in little things around the grave or larger things in the surrounding landscape or even in sun, wind and sky, or, for the brave, the tuning in with the earth-and-soil story: all may contribute to healing and a re-learning of the world (Attig 1996). It remains a question, however, to what extent grieving can ever be phased. Much about death remains ambiguous. Reality, especially in the face of death, can often be taken merely 'by degrees': 'the task of reality acceptance is never completed' (Winnicott 1971: 13). For some, whose life feels 'fulfilled', dying itself may be consolation. For others, death remains an enemy, whatever stories are told to soften it. And however enchanted or even sacred such a site may look to those who have tuned in with the greater perspective, it remains an emotionally challenging 'necro-scape' and 'necro-geography' (Kniffen 1967).

Moreover, to most, tree symbolism (Cloke and Pawson 2008; Davies 1988; Rival 1998) may be much more attractive – and provide a more universal consolation-scape – than the soilscape with its layers of compressed dead matter. It needs courage and vision to share the perspective of the soil, however alive these layers may be, a truly 'wood wide web' (Simard et al. 1997). Even today, in cemeteries and crematoria, there is a cultural preference for upwardly mobile symbols like cherubs, birds, butterflies, balloons, kites, and of course trees. As objects, trees seem sessile, passive, pliably responsive to human acts and needs. And with the great age they are capable of reaching, they express something ineffable and awe-inspiring about the continuity of life on earth (Battles 2017). Whereas all over the world mature trees are easily recognised as instances of the sublime – surrounding us, enfolding us, embowering us – most people abhor the soil and have a deeply ingrained fear of the dark, dense, oppressing feel of the earth. As far as narratives go, the soil story, however poetic to some, and however fascinatingly related to by Peter Wohlleben, may not win the consolation prize right away.[2]

---

2   Methodological justification: My own set of empirical data underlying this article qualifies as 'ethnographic monitoring' (Van der Aa and Blommaert 2015) as it is based on the ongoing visits to three natural burial sites in my direct vicinity. They are located by the southeastern border between the Netherlands and Germany: Weverslo (between Venray and Deurne), Venlo-Maasbree, and Bergerbos (St. Odiliënpeel). Regular observations over the course of years gave me insights into the various stages of particular graves as well as of the sites in general. Visual data of the graves and the ritualised behaviour of visitors were corroborated by informal conversations with various types of stakeholders, as well as by in-depth interviews with professional key persons. I have visited most of the other natural burial sites in the Netherlands (around 20) only once. During a lifetime of travel around the globe I informally looked around at many other burial sites. The discourse on nature/Nature I address in this article is partly shaped and informed by my ongoing research on rituals around sacred trees in India (2005; 2018), religiously inspired environmentalism (2009), dispersal of ashes in 'riverscapes' (2017) and 'seascapes' (2014b) and the ecological 'costs' of outdoor Hindu cremation rituals in South Asia (2016).

References

Albery, N., Elliot, G., and Elliot, J., eds. 1997. *The New Natural Death Handbook.* London: Vintage/Ebury.

Ariès, P. 1981. *The Hour of Our Death.* London: Allen Lane.

Ashworth, G. J. and Graham, B. 2005. *Senses of Place, Senses of Time.* Aldershot: Ashgate.

Asprem, E. 2013. *The Problem of Disenchantment: Scientific Naturalism and Esoteric Discourse, 1900–1939.* Amsterdam: University of Amsterdam PhD thesis. htttp://dare.uva.nl/record/436829/

Attig, T. 1996. *How We Grieve: Relearning the World.* New York: Oxford University Press.

Augé, M. 1995. *Non-places: Introduction to an Anthropology of Supermodernity.* London: Verso.

Barthes, R. 1972. *Mythologies.* London: Paladin. Original French edition (Paris: Editions du Seuil, 1957) translated by Annette Lavers.

Basmajian, C. and Coutts, C. 2010. Planning for the disposal of the dead. *Journal of the American Planning Association* 76/3, 305–317.

Battles, M. 2017. *Tree. Bloomsbury Object Lessons.* London: Bloomsbury.

Bender, B. 2006. Place and landscape, in *Handbook of Material Culture*, edited by C. Tilley, W. Keane, S. Küchler, M. Rowlands, and P. Spyer. London/Thousand Oaks, CA: Sage, 303–315.

Bucknill, P. J. R. 1915. Cremation: the only rational means of the disposal of the dead. *Journal of the Royal Society for the Promotion of Health* 36/1, 54–56.

Canning, L. and Szmigin, I. 2010. Death and disposal: the universal, environmental dilemma. *Journal of Marketing Management* 26/11&12, 1129–1142.

Castree, N. 2003. Commodifying what nature? *Progress in Human Geography* 27/2, 273–297.

Clayden, A. 2013. Review of *Natural Burial: Traditional/Secular Spiritualities and Funeral Innovation. Mortality* 18/3, 324–325.

Clayden, A. and Dixon, K. 2007. Woodland burial: memorial arboretum versus natural native woodland? *Mortality* 12/3, 240–260.

Clayden, A., Hockey, J., and Powell, M. 2010. Natural burial: the de-materializing of death, in *The Matter of Death: Space, Place and Materiality*, edited by J. Hockey, C. Komaromy, and K. Woodthorpe. Basingstoke/New York: Palgrave MacMillan, 148–164.

Cloke, P. and Pawson, E. 2008. Memorial trees and treescape memories. *Environment and Planning (D: Society and Space)* 26/1, 107–122.

Davidson, J., Bondi, L. and Smith, N., eds. 2005. *Emotional Geographies.* Burlington, VT: Ashgate.

Davies, D. J. 1988. The evocative symbolism of trees, in *The Iconography of Landscape: Essays on the Symbolic Representation, Design and Use of Past Environments*, edited by D. E. Cosgrove and S. Daniels. Cambridge: Cambridge University Press, 32–42.

Davies, D. J. 2002. *Death, Ritual and Belief: The Rhetoric of Funerary Rites.* London: Continuum.

Davies, D. J. 2005. *A Brief History of Death.* Malden, MA: Blackwell.

Davies, D. and Rumble, H. 2012. *Natural Burial: Traditional –Secular Spiritualities and Funeral Innovation.* London/New York: Continuum.

De Haas, W. and De Vries, B. 2013. *Natuurbegraafplaatsen in Nederland: Landelijke Inventarisatie 2013.* Wageningen: Alterra.

Eisenstadt, S. 1999. Multiple modernities in an age of globalization. *Canadian Journal of Sociology* 24/2, 283–295.

Feagan, R. 2007. Death to life: towards my green burial. *Ethics, Place, and Environment* 10/2, 157–175.

Fox, W. 1990. *Toward a Transpersonal Ecology: Developing New Foundations for Environmentalism.* Boston/London: Shambhala.

Fox, W. 2006. *A Theory of General Ethics: Human Relationships, Nature, and the Built Environment.* Cambridge, MA: MIT Press.

Fromm, H. 2009. *The Nature of Being Human: From Environmentalism to Consciousness.* Baltimore: The Johns Hopkins University Press.

Gesler, W. 1992. Therapeutic landscapes: medical issues in light of the new cultural geography. *Soc.Sci.Med.* 34/7, 735–746.

Giannachi, G. and Stewart, N. eds. 2005. *Performing Nature: Explorations in Ecology and the Arts.* Bern: Peter Lang.

Gibson, J. W. 2009. *A Re-enchanted World: The Quest for a New Kinship with Nature.* New York: Metropolitan Books.

Groote, P. and Klaassens, M. 2010. Bergerbos: a community of the bereaved. *Annals of the University of Alba Iulia*, History Special Issue 2, 311–327.

Harari, Y. N. 2016. *Homo Deus: A Brief History of Tomorrow.* London: Vintage.

Harris, M. 2007. *Grave Matters: A Journey through the Modern Funeral Industry to a Natural Way of Burial.* New York: Scribner.

Heise, U. K. 2008. *Sense of Place and Sense of Planet: The Environmental Imagination of the Global.* New York: Oxford University Press.

Hockey, J., Komaromy, C. and Woodthorpe, K., eds. 2010. *The Matter of Death: Space, Place and Materiality.* Basingstoke/New York: Palgrave MacMillan.

Janowski, M., and Ingold, T., eds. 2012. *Imagining Landscapes: Past, Present and Future.* Farnham: Ashgate.

Kaufman, M. 1999. Dust to dust? A greedy death industry prevents our return to nature. *Conscious Choice,* 8–9.

Kearl, M. C. 2010. The proliferation of postselves in American civic and popular cultures. *Mortality* 15/1, 47–63.

Kearns, R. A. and Gesler, W. M. 1998. *Putting Health into Place: Landscape, Identity, and Well-Being.* Syracuse/New York: Syracuse University Press.

Klaassens, M. 2011. *Final Places: Geographies of Death and Remembrance in the Netherlands.* Amsterdam: Rozenberg.

Klass, D., Silverman, P. R. and Nickman, S. L., eds. 1996. *Continuing Bonds: New Understandings of Grief.* London/Washington, DC: Taylor & Francis.

Klass, D. 2006. Grief, religion, and spirituality, in *Death and Religion in a Changing World*, edited by K. Garces-Foley. Armonk, NY: M.E. Sharpe, 283–304.

Kniffen, F. 1967. Necro-geography in the United States. *Geographical Review* 57/3, 426–427.

Kramer, K. 1988. *The Sacred Art of Dying: How World Religions Understand Death.* New York: Paulist Press.

Kunkel, A. D. and Dennis, M. R. 2003. Grief consolation in eulogy rhetoric: an integrative framework. *Death Studies* 27/1, 1–38.

Leopold, A. 1970. *A Sand County Almanac: With Essays on Conservation from Round River.* New York: Ballantine Books.

Lifton, R. J. 1979. *The Broken Connection.* New York: Simon & Schuster.

Maddrell, A. and Sidaway, J. D. 2010. Introduction: Bringing a Spatial Lens to Death, Dying, Mourning and Remembrance, in *Deathscapes: Spaces for Death, Dying, Mourning and Remembrance,* edited by A. Maddrell and J. D. Sidaway. Farnham: Ashgate, 1–16.

Naes, A. 1973. The shallow and the deep, long-range ecology movement: a summary. *Inquiry* 16, 95–100.

Nolan, S. 2013. Earth from an alien's eye view: how our planet looks from a different perspective far, far away. *Mail Online,* updated online edition July 30. www.dailymail.co.uk/sciencetech/article-2380085/

Nugteren, A. 2005. *Belief, Bounty, and Beauty: Rituals around Sacred Trees in India.* Leiden/Boston: Brill.

Nugteren, A. 2009. From cosmos to commodity … and back again: a critique of Hindu environmental rhetoric in educational programs, in *Religion and Sustainable Development: Opportunities and Challenges,* edited by C. De Pater and I. Dankelman. Münster: LIT, 159–168.

Nugteren, A. 2014a. Troost en 'troosteloosheid' op natuurbegraafplaatsen. *Nederlands Theologisch Tijdschrift* 68/1–2, 83–100.

Nugteren, A. 2014b. Landscapes that save, seascapes that soothe: places, traces, and intended oblivion. Powerpoint presentation, symposium Sense of Place, Waddenacademie, Terschelling, June 11–12, 2014. Accessible through https://www.waddenacademie.nl (symposium June 2014, presentations).

Nugteren, A. 2015a. Spaces, places, traces: an afterlife for the body in natural burial processes, in *The Study of Culture through the Lens of Ritual,* edited by L. Sparks and P. Post. Groningen: *Netherlands Studies in Ritual and Liturgy,* 15, 163–177.

Nugteren, A. 2015b. A darker shade of green? An inquiry into growing preferences for natural burial, in *Religious Diversity Today: Experiencing Religion in the Contemporary World,* edited by A. Panagakos. Santa Barbara, CA: ABC/CLIO, 2:111–134.

Nugteren, A. 2016. Woods, water, and waste: material aspects of mortuary practices in South Asia, in *Roots of Wisdom, Branches of Devotion: Plant Life in South Asian Traditions,* edited by F. Ferrari and T. Daehnhardt. London: Equinox, 118–141.

Nugteren, A. 2017. Through fire: creative aspects of sacrificial rituals in the Vedic-Hindu continuum, in *Sacrifice in Modernity, Ritual, Identity: From Nationalism and Non-Violence to Health Care and Harry Potter,* edited by J. Duyndam, A.-M. Korte, and M. Poorthuis. Leiden/Boston: Brill, 109–131.

Nugteren, A. 2018. Sacred trees, groves, and forests, in *Oxford Bibliographies in Hinduism.* Editor-in-chief Tracy Coleman. New York: Oxford University Press, 1–83.

Rival, L. (ed.) 1998. *The Social Life of Trees: Anthropological Perspectives on Tree Symbolism.* Oxford/New York: Berg.

Rose, G. 2007. *Visual Methodologies: An Introduction to the Interpretation of Visual Materials.* London: Sage.

Rumble, H. 2010. *Giving Something Back: A Case Study of Woodland Burial and Human Experience at Barton Glebe.* Durham University PhD thesis. http://etheses.dur.ac.uk/67911/Rumble_thesis.pdf

Schade, T. L. 2011. *The Green Cemetery in America: Plant a Tree on Me.* MA thesis The Evergreen State College, Olympia, WA.

Schama, S. 1995. *Landscape and Memory.* London: HarperCollins.

Sheldrake, R. 1991. *The Rebirth of Nature: The Greening of Science and God.* New York: Bantam Books.

Simard, S. W., Perry, D. A., Jones, M. D., Durall, D. D. and Molina, R. 1997. Net transfer of carbon between the tree species with shared ectomycorrhizal fungi. *Nature* 388, 579–582.

Soulé, M. E. 1995. *Reinventing Nature? Responses to Postmodern Deconstruction.* Washington, DC: Island Press.

Smith, N., Davidson, J., Cameron, L. and Bondi, L., eds. 2009. *Emotion, Place, and Culture.* Burlington, VT: Ashgate.

Szerszynski, B., Heim, W. and Waterton, C., eds. 2003. *Nature Performed: Environment, Culture and Performance.* Oxford: Blackwell.

Taylor, B. 2010. *Dark Green Religion: Nature, Spirituality and the Planetary Future.* Berkeley/Los Angeles/London: University of California Press.

Van der Aa, J., and Blommaert, J. 2015. Ethnographic monitoring and the study of complexity. *Tilburg Papers in Culture Studies* 123. https://www.tilburguniversity.edu/upload/24266e94-2d00-41fc-b488-60e3845fc383_TPCS_123_VdrAa-Blommaert.pdf/

Van Tongeren, L. 2004. Individualizing ritual: the personal dimension in funeral liturgy. *Worship* 78/2, 117–138.

Venbrux, E., Peelen, J. and Altena, M. 2009. Going Dutch: individualisation, secularisation and changes in death rites. *Mortality* 14/2, 97–101.

Venbrux, E., Quartier, Th., Venhorst, C., and Mathijssen, B., eds. 2013. *Changing European Death Ways.* Münster: LIT.

Walter, T. 1996. A new model of grief: bereavement and biography. *Mortality* 1, 7–25.

Walter, T. and Gittings, C. 2010. 'What will the neighbours say? Reactions to field and garden burial,' in *The Matter of Death: Space, Place and Materiality*, edited by J. Hockey, C. Komaromy, and K. Woodthorpe. Basingstoke/New York: Palgrave MacMillan, 165–177.

West, K. 2010. *A Guide to Natural Burial.* London: Thomson Reuters.

White, F. 1987. *The Overview Effect: Space Exploration and Human Evolution.* Boston: Houghton Mifflin.

Wienrich, S. and Speyer, J. 2003. *The Natural Death Handbook.* London: Rider.

Williams, A. (ed.) 1999. *Therapeutic Landscapes: The Dynamic between Place and Wellness.* Lanham/New York/Oxford: University Press of America.

Willers, W. B. 1999. *Unmanaged Landscapes: Voices for Untamed Nature.* Washington, DC: Island Press.

Winnicott, D. W. 1971. *Playing and Reality.* New York: Basic Books.

Wohlleben, P. 2017. *The Hidden Life of Trees: What They Feel, How They Communicate: Discoveries from a Secret World.* Trans. Jane Billinghurst. London: William Collins.

4 The crematorium as a ritual and musical consolationscape

Martin J. M. Hoondert

Introduction

The Tilburg crematorium was officially opened in 1986. In the more than 30 years of its existence, it has developed into an important centre of mourning and memorial rituals (Hoondert 2015). During its construction, attention was paid not only to the architectural design, but also to music and sound. A modern sound system was integrated into the building. From the beginning it was possible to choose tracks from the playlist ('Muziekboek'), to supply tracks from one's own CDs that were not on the playlist, or to sing and play live music (de Leeuw 2009: 64, 81). The crematorium's playlist, which is now an online archive for families of the deceased, contains well over 4000 tracks. For the crematorium's employees, the amount of time they need to invest in a careful handling of the families' musical wishes has increased noticeably. The manager compares the most recent use of music with that of the early days:

> In your regular funeral service there used to be three pieces of music. And one of these would be the classic 'Waarheen, waarvoor?' by Mieke Telkamp (adaptation of the well-known song 'Amazing Grace', about the meaning of life [MH]). In between, there would be two speakers or texts being read. Now there's no end to what people can choose: eleven musical pieces and no speaker, live music. Sometimes relatives sing or perform music. There's more work involved in the music than there used to be.
>
> (de Leeuw 2009: 96)

Owing to the personalisation of funeral rituals, music has become a dominant element, especially in a crematorium which is very well equipped with professional sound systems (Parsons 2012). It is surprising, therefore, that in the literature the theme of 'music and cremation' should have received so little attention. Thus, in the *Encyclopedia of Cremation* (Davies 2005a), music is not a separate entry but is subsumed in certain national developments (for example in Belgium, p. 90), or in cultural developments as secularisation (p. 374). This matches the general lack of attention to music as part of a ritual or a ritualised context. Although scholars do hear the music when carrying out fieldwork into all kinds of ritual, the description, analysis and interpretation of music *in* and *as* ritual is often absent. For example, in Catherine Bell's widely used handbook of ritual studies, *Ritual: Perspectives and Dimensions*, music is absent from the index (Bell 1997). Likewise, Ronald Grimes, one of the founding fathers of ritual studies, pays hardly any attention to music in his publications. In his autobiographical book *Marrying and Burying*, he writes some moving chapters about the death and burial of his son, his grandmother, and his parents, but it is only in the description of his mother's burial that music is mentioned at all, and then only in one short paragraph (Grimes 1995: 136). In his latest book about the craft of ritual studies, Grimes describes the role of music in ritual in a seven-page section (Grimes 2014: 217–223), introducing the work of the cultural musicologist Christopher Small (Grimes refers to Small 1998). Unfortunately, Grimes's approach to music remains abstract and is not very innovative.

In this chapter, music will be the focus of attention, as music embedded in the physical space of the crematorium and the metaphorical space of the ritual (Hoondert and Bruin-Mollenhorst 2016). I will explore the crematorium, not from an architectural perspective but from a musical and musical-ritual perspective. The focus will be on the function and the effect of music as part of the crematorium ritual in relation to consolation. More specifically, the research questions are: What is the music that we hear in the crematorium? How does that music function in relation to the crematorium as a building and to the ritual taking place within it? Is music capable of offering consolation? These questions will be dealt with by analyzing the playlist of the Tilburg crematorium and by ethnographic research on the use of music in cremation rituals. I shall start by focusing on the relation between music and the crematorium as a ritual space. After that, I shall explore the complex theme of music and consolation. Finally, I shall present some results of my research on music, consolation and the cremation ritual.

Music and the cremation ritual

We can approach the relation between music, the crematorium and the cremation ritual in three different ways: (1) music is framed; (2) music is a social activity; and (3) music produces a 'musical place'. Each of these sheds a different light on the same issue.

To begin with, music as part of the cremation ritual is not music that is listened to neutrally. It is framed by the circumstances, by the cremation ritual and the crematorium as a ritual place, and it is linked to the deceased and the families who loved them. The circumstances cause the music to be listened to as an expression of grief and mourning, or as a remembrance of or memento to the life of the deceased. Framing is an important dimension of both ritual and place. According to Gregory Bateson (1904–1980), a frame is a form of meta-communication (Bell 1997: 74). Thus, it is not the music (as sound) itself that points to the deceased but the frame of this particular ritual in this particular place and circumstances that causes us to link the music to the life of the deceased. Actually, a long discussion related to the field of the philosophy of music regarding the referential potentiality of music should precede this paragraph on framing. However, I refrain from this discussion, stating simply that music is about music and needs to be framed to have that referential potentiality. Constantijn Koopman (Koopman 1999; Koopman and Davies 2001) and Peter Kivy (1990) have written more eloquently on this topic than I can do.

Second, in a crematorium, both listening to music and performing it are social activities. Music as part of the ritual has a certain impact on the participants: they become listeners, whether they like the music or not. The music can invite people to engage in the ritual, but it might also 'frighten them off' or deter them, as it were. The music joins together both music-lovers and music-haters. The performance of music as a social activity implies a complex meaning-making process: meanings are designated by framing, but the participants in the ritual remain active as meaning-makers. So, meanings are also appropriated, a process that puts the importance of framing into perspective.

Third, the performance of music is a spatial practice (Lefebvre 2004), turning the space where it is performed into a 'musical place'. We tend to speak of music as an art form that unfolds in time. However, music also creates a place. The activity of making music does something to the space in which it is performed and to those present in that space. On the one hand, this is an acoustic and measurable phenomenon: the musical place is shaped by the objective aspects of the sound produced. On the other hand, the sound also influences those present in the musical place by appealing to the body and to associations, memories, meanings and emotions. These are the subjective aspects of the sound, which can be physiological, psycho-acoustic and aesthetic dimensions (Feiereisen and Merley Hill 2012; Hoondert 2012). The music fills the place and is, in a sense, inescapable for all those who are present. As an architectural place, the crematorium is relatively stable and unchanging. As a musical place, it changes according to the performance of the music, in keeping with the basic emotions that are communicated (Johnson-Laird and Oatley 2008).

Interference of places: Building, ritual and music

By applying a spatial perspective to the music in a cremation ritual, we might say that sound and music, which create a place in a metaphorical sense, interact with two other places: the physical place of the crematorium (the building) and the ritual place of the cremation. The crematorium as a building is a physical place associated with death and remembrance; ritual and music belong to the category of embodied-psychological places of grief and mourning (Maddrell 2016). Both ritual and music are activities in which the body is involved (Nugteren 2013). 'The body can be seen (…) as a site of embodiment, of identity, experience, performance. It is a space where things happen' (Maddrell 2016: 176). Although rituals are to a great extent verbal, they reside in the body. Ronald Grimes firmly stated: 'No body, no ritual' (Grimes 2014: 306).

Much research has been carried out into the crematorium as a building (Grainger 2005; Klaassens and Groote 2012, 2014). In a master's thesis at the University of Tilburg, Laura Cramwinckel compared six crematoria in the Netherlands from architectural and religious studies perspectives (Cramwinckel 2011). The comparison showed that when crematoria were being constructed or re-styled in the years after 2000, there was an emphasis on symbolism and ritual. The sober and functional building style that characterised crematoria prior to 2000 was abandoned. There is a tendency towards both the resacralisation and the ritualisation of the crematorium (Cramwinckel 2011: 60–68). When designing new crematoria, architects strive to realise the sacral character of the building and the ritual that takes place there. They work with fragmented light-fall and shadow, contrasts between openness and closedness, zones and thresholds, to emphasise the liminal character of the ritual. When crematoria are re-styled, the emphasis is on warmth and hospitality, on consolation and beauty. Cramwinckel writes:

> The decor of the auditorium is far from 'commonplace'. In relation to the affective aspect, the designers speak of creating 'consoling' atmospheres, spaces which inspire, allow one to 'dream away' or reflect. They are clearly defined spaces with an illusory character. For the material execution the stylists, as well as the architects, make use of the evocative qualities of beauty, art and applied design. In order to allow for interpretations from different ideological backgrounds, consciously (and subconsciously) fundamental-sacral visual language and universal metaphors are used, such as nature, light, harmony, life and hope.
>
> (Cramwinckel 2011: 63)

Besides the physical place of the crematorium, the ritual place plays an important role in the way we experience music as part of the crematorium ritual. As with music as spatial practice, I use the phrase 'ritual place' as a metaphor. The ritual acts as a 'place' which the participants enter and which imposes certain behaviour and frames. Architecture (building) and ritual can complement each other, but they can also get in each other's way.

The music as part of the cremation ritual takes place in the physical place of the crematorium (the building) and in the ritual place.

The three above-mentioned places are closely related but they are not one, as is shown in Figure 4.1. The crematorium as a building is more than a ritual place: it also has offices, a furnace or committal room, and a meeting-room for giving and receiving condolences. The committal room is separated from the auditorium/chapel and constitutes a second ritual place in the sequence of ritual stages in the crematorium. The committal room contains not only the furnace but also installations that, inter alia, prevent the emission of toxic fumes (such as mercury vapour). It is first and foremost a technical and functional place which is made accessible for the bereaved to witness the coffin being placed in the cremator. The meeting room is the place for the

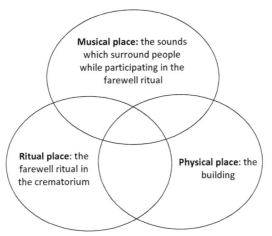

Figure 4.1 Overlapping physical, ritual and musical places

third ritual in the sequence of rituals in the crematorium: a reception, the possibility to give and receive condolences, the opportunity to eat and drink while the cremation takes place (Bryant 2003: 624).

The crematorium as a building is determined by its spatial practices (Lefebvre 2004). The place changes during use; it is both dynamic and fluid: lectures can be given in the auditorium/chapel, while during the evening it can be used for choir practices. The place is not something that *is* but something that is *created*; it is a matter of ontogenesis through what Dodge and Kitchin call 'transduction': 'the constant making anew of a domain in reiterative and transformative practices' (Dodge and Kitchin 2005: 162). Although the crematorium is a cultural and social place for saying farewell to the deceased and for their disposal, the physical place needs to be appropriated through personal and collective practices of meaning-making. Rituals and music render the crematorium a meaningful place.

The ritual space consists of more than music alone. The ceremony in the auditorium/chapel is brought about by the interaction between physical space (light, colour, physical quality, connection with the outside world), words, actions and music; it is also related to preceding rituals (e.g. at home or in church) and subsequent rituals (e.g. transfer to the furnace, the funeral reception where condolences can be offered). The ritual space can cross the boundaries of the here and now of the ritual through the use of photo presentations and videos of the deceased and through the online participation of family and friends at a greater distance afforded by a webcam and live streaming of the ceremony (Maddrell 2016: 178–179).

The music fulfils a ritual function but does not correspond with it: the music not only has meaning within the context of the cremation ritual; it also carries a wide variety of meanings. It evokes memories of other contexts in which it has been heard (Hoondert 2014), it represents a culture and can either connect the participants with or distract them from the ritual. Participation in music is more than just the acoustic functioning of our ears. We find ourselves part of the music that surrounds us. This musical experience is physical, but it is also social and cultural (Small 1998). We do not receive music passively; we appropriate it actively, though not necessarily consciously. It is through musical appropriation that music affords 'shifting emotional states, remembering, concentrating, creating or clarifying identity, enhancing personal relationships, socialising, or finding "the spiritual"' (Ansdell 2014: 38). In relation to purely musical activities other, non-musical things (emotions, memories, etc.) take shape. As such, music creates an embodied-psychological place to abide and wander around.

Music and consolation

One of the non-musical processes that might take shape through participating in music during a cremation ritual is consolation. I consider 'being consoled' as a subjective experience which, in this case, is evoked by the live performance of music or by recorded music being played in the context of a cremation ritual. There is a causal link between the perception of a piece of music and the ensuing consoling experience. To understand this causal link we have to explore the concept of consolation, a concept which, so far as I have been able to find out, appears to be surprisingly underdeveloped.

Alexander Stein defined consolation from a psychoanalytical perspective as 'to imply something (or someone) that provides or offers reparation, repair, or relief from discomfort' (Stein 2004: 803). Stein continued by stating that 'the underpinnings of consolation are closely related to the capacity for empathy and are perhaps even predominantly derivative of it'. In empathy, we share the experience of others as our own. Stein terms this perceptual constellation 'empathic imagination' (Stein 2004: 803). In the sphere of bereavement and mourning, consolation can provide a significant counterpoint to the reactions of disconnection and painful feelings of desolation or isolation. Consolation is thus, ideally, restitutive and reconstructive (Stein 2004: 805).

Focusing on music as sound, and thus on the aural, wordless experience of music, we can distinguish three dimensions that contribute to the consoling experience of music: the ability of music to express basic emotions, the experience of time that music evokes, and the ability of music to foster self-empathy. In what follows, I will elaborate on these three dimensions.

According to Johnson-Laird and Oatley, music creates emotions in a mimetic way: 'It mimics the main characteristics of emotional behavior, speech, and thought' (Johnson-Laird and Oatley 2008: 107). This concerns only the basic emotions of happiness, sadness, anxiety, and anger, which are communicated through a small set of nonverbal signals that do not require working memory for their interpretation. These hypotheses are part of a communicative theory of emotions (Johnson-Laird and Oatley 2008), which I consider a convincing and workable theory. Music is capable of conveying sadness and grief-related affect by signals such as a slow tempo, slightly discordant harmonies and a muted sound (Johnson-Laird and Oatley 2008: 107). As such, music can function as 'an externalization of an intolerably overwhelming, incomprehensible, or crushing internal state' (Stein 2004: 808). It helps the relatives not only to express their grief, but also to practice handling that grief. This externalisation of grief is less threatening than the internally felt grief (Honing 2012: 26–28), and music, as such, has a comforting and consoling effect on the listener.

Second, music is able to transform our perceptual and sensory experience of time. Music consists of sounding tones. These tones are not independent, they are interrelated and form part of a system of relations. The relations between the tones give each tone a dynamic quality as a result of which the tones are in constant motion, always moving as they are towards the next tone. It is this dynamic quality that makes a tone a musical fact (rather than a mere sound). When we are listening to music we are sharing in these dynamics and this transforms our experience of time. Music unfolds in time, but the relations between the tones, the expected rises and falls and those actually taking place, make us experience past, present and future all at the same time. In the 'now' of the music we hear the tones that have already sounded and we anticipate the tones yet to come. We hear the tone in the relational network of tones, but what we are in fact hearing is always a 'now' in which past and future reverberate. To put it differently: in the 'now' of the music we experience time in its broadest form. The musician and liturgist Kathleen Harmon wrote: 'Musical hearing is (…) presence to and participation in the completeness of time in every present moment' (Harmon 2008: 33). This participation in the completeness of time relieves or diminishes feelings of pain 'by providing an illusory response ensconced in rhythm and sound to the dominant wish of the bereaved – reunion with the lost object' (Stein 2004: 807).

Third, music has the potential to foster self-empathy. To understand this, we need to give some thought to the way music is perceived as meaningful. Music, understood here as sound, is without propositional content. So, to theorise on music and meaning, we cannot revert to the language paradigm as many philosophers of music do. The issue of music and meaning has been summarised by Constantijn Koopman and Stephen Davies in two articles and I am relying on their insights here (Koopman 1999; Koopman and Davies 2001). One of their approaches to the issue is a comparison between the meaning that music has for the listener and the meaning that other people have for us. The philosopher Roger Scruton, among others, characterised our response to music as one of empathy ('Einfühlung'): 'you are the music while the music lasts' (Scruton 1999: 364). Approaching the issue from a psychological perspective, Stein stated: 'music listening invokes an imaginative resurrection of an internalised other […]. The work of music can thus itself be conceived as a responsive object.' Referring to a book by the psychoanalyst Christopher Bollas (*The Shadow of the Object*, 1987), Stein continued, speculatively: 'This, as Bollas suggests, is a "pre-verbal, essentially, pre-representational registration of the mother's presence" (p. 39) and will thus facilitate a restitutive transformation of internal experience and affect; we feel held, understood, consoled' (Stein 2004: 807).

From a purely musical perspective, the cultural musicologist Christopher Small (1927–2011) reached the same conclusion, using the verb 'musicking' instead of the noun 'music' (Small 1998). By doing this, he intended to shift the focus from music as an object to music as an activity. Moreover, he emphasised how music cannot exist without performance. In order to stress this performative approach of music, Small used the verb 'musicking' to address musical performances as an encounter between humans. Through the performance, relationships between the sounds are brought into existence. These sound relationships articulate tensions and relaxations, paralleling the development of human relationships. Small wrote: 'all musicking can be thought of as a process of storytelling, in which we tell ourselves a story about our relationships' (Small 1998: 139). The storytelling process is carried out by means of the language of gesture: there are no nouns and gestures are always in the present tense. This means that the universe of relationships is much richer than can be dealt with in words. In the virtual world of the performance we may experience relationships as being restored. This is a cause for elation and consolation, but also for melancholy, because we recognise that those restored relationships exist only in the virtual world of the performance and not in the everyday world to which we must return after the performance is over (Small 1998: 130–143). It seems that music dances between presence and the presence of something or someone absent (Maddrell 2013).

Crematorium, music and consolation

The experience of music as sound, which can be a consoling experience, is only one of the factors that makes the cremation ritual a meaningful ritual, which helps the family and friends to say farewell to their loved one. In this section, I shall pursue the question of music and consolation, and the crematorium as a consolationscape, taking into account the complex reality of music, lyrics, ritual and place. In order to do this, I shall first analyse the music of ten cremation rituals, and then I shall examine the music embedded in the ritual context of one particular cremation ritual.

As a first approach to the complex reality of music, lyrics, ritual and place I shall analyse the music of ten cremation rituals that took place in the first half of January 2013 at the Tilburg crematorium. This selection comprises the first ten rituals in which tracks from the playlist were relayed. Rituals involving people's own recordings were excluded. My reason is a practical one: the titles and performers of the tracks on the playlist can be traced, the data of people's own recordings are unknown to me. The selection comprises a total of 45 tracks, 28 of which belong to the genre of popular music (pop, jazz, folk, easy listening), and 17 to the genre of classical music. Seven of these 17 classical tracks are instrumental. For six cremations a version of the *Ave Maria* (J S Bach/Gounod or Schubert) was chosen. *Con te partiro* (Italian text) or *Time to say goodbye* (English text), sung by Andrea Bocelli and Sarah Brightman, was played on five occasions, *Die letzte Rose* (an opera classic, composed by Friedrich von Flotow), *Et les oiseaux chantaient* (an instrumental piece performed by the band 'Sweet people') and *The Rose* (English text) or *De roos* (Dutch text; a song composed by Amanda McBroom which became famous through the performance by Bette Midler in 1979) were all played twice.

Only a few tracks are explicitly connected with death and farewell: *Con te partiro* and *Time to say goodbye*, the *Introit* from the Gregorian Requiem Mass and the Dutch-language tracks *Trein naar niemandsland* (Train to no man's land), *Je naam in de sterren* (Your name in the stars) and *Zo zal het zijn* (That's how it will be). Religious notions are not well represented in the music of the ten cremations that were examined.

The 'Ave Maria' is the most frequently heard song in the ten analysed cremation rituals. It is also the most frequent title on the Tilburg playlist, in versions composed by J S Bach/Gounod, Schubert, Caccini, Mendelssohn, or sung as a Gregorian chant. The title occurs 82 times in many different versions and performances, both vocal and instrumental. The relation between Marian devotion and the commemoration of the dead has been investigated by Catrien Notermans and Heleen Kommers. They explored the meaning Mary has for pilgrims to Lourdes:

> In Dutch society, churches have been gradually losing the role of helping people to venerate their deceased loved ones. (…) Now that the churches no longer connect people to their past and their deceased relatives, it is Mary who provides continuity rather than the church. Being in contact with Mary is being in contact with a cherished past to which the deceased family members belong. Mary helps with sustaining the mutual dependence between the living and the dead: the dead need to be remembered by the living and the living need the dead in order to feel wholeness and continuity in life. In this way, Mary creates intergenerational continuity in times of discontinuity, multiple losses, and social and physical disempowerment.
>
> (Notermans and Kommers 2012: 13)

In the case of the 'Ave Maria', the effect of the music as a responsive object, to use Stein's terminology, is doubled: the prayer to Mary, or rather the image of Mary evoked by the music and the lyrics, and the music itself, to quote Stein again, make the listener feel held, understood and consoled. Besides this, the music of J S Bach/Gounod and Schubert's 'Ave Maria' not only belongs to the ritual-musical repertoire of funerals and cremations, but also to that of weddings. These two compositions are so well-known that they can be said to be part of the ritual-musical toolkit readily available to everyone.

Are there features that characterise the music of the 45 tracks, taken from the ten cremation rituals that were analysed? With a certain amount of caution, we can conclude that a musical repertoire for cremation rituals is outlined here. It consists of the *Ave Maria, Time to say goodbye*, and slow, romantic music, of which Grieg's *Morgenstimmung* and *The Rose* are good examples. Of the 45 tracks of the above-mentioned ten cremation rituals there are only four compositions that do not fit this profile: *Blueberry Hill* by Fats Domino, *Friends will be friends* and *Bohemian Rhapsody* by Queen, and Handel's *Hallelujah Chorus*. We can describe this type of romantic music as 'comforting music' creating a mild level arousal (Mollenhorst et al. 2016). Nikki Rickard, in her research on music, emotion and arousal, speaks of 'relaxing music' and concludes that this kind of music significantly reduces skin conductance and the number of chills, which are physiological indications of an emotional response (Rickard 2004). It appears that the main effect of the musical repertory of the analysed cremation rituals is that it channels emotions (Lukken 2005: 54–73). The romantic, comfort music one often hears at cremations – in general – has a slow tempo, smooth rhythms and relatively few dissonant harmonies. The absence of dissonance makes that, according to Johnson-Laird and Oatley, there is no question of the mimesis of the basic emotion 'sadness'. The music does not express grief or sorrow, but rather it appears to have a therapeutic effect and to be aimed at bringing about a transformation into a different mood (Robinson 2005: 393). Here the three 'places' of the cremation ritual converge. The mood or state of mind that is evoked by the musical place, corresponds with what both the physical and the ritual places seem to evoke. The physical place (the building of the crematorium) appears to evoke 'softness': the warm colours of the furniture, the dimmed lighting and view of nature are reflected in the music. And the functions of the ritual possibly require a musical style that does not arouse strong emotions. My hypothesis is therefore that the romantic style of the music that is part of the cremation ritual, which in itself is not more capable of bringing about a feeling of consolation than other styles, is strongly determined by the physical and ritual space.

To explore the question of music and consolation somewhat further I will describe and analyse the cremation of Mrs W. She died at the age of 79, leaving behind her husband, her son and his wife and a still unborn grandchild. The cremation ritual was led by a funeral celebrant, who was well acquainted with Mrs W's life-history. The order of the ritual, which lasted three-quarters of an hour, is as follows:

1   Entrance of family and friends, accompanied by quiet classical music, unknown to me
2   Welcome and poem by the funeral celebrant
3   Music: *Goodbye my love*, sung by Marianne Weber
4   Speaker outlines the course of Mrs W's life
5   Music: *My way*, sung by Frank Sinatra
6   Closing speech, 'Hail Mary' and poem
7   Parting with music from a barrel organ

The funeral celebrant was the only one who spoke. At the end, she invited all those present to join her in praying the 'Hail Mary', but only a few of them participated. In my opinion, the song *Goodbye my love* expressed the feelings of the husband, and the son and daughter-in-law in particular. *My way* was chosen to express the way Mrs W. lived her life. The music of the barrel organ was chosen because Mrs W. used to go out and dance to the music of a barrel organ every Sunday afternoon. We can characterise *Goodbye my love* as a farewell song, but *My way* and the music of the barrel organ are more or less about keeping alive the memory of the deceased. These two pieces of music were used in service of the résumé of Mrs W's life and they are an expression of the tendency towards a personal memorial ritual characterised by biographical information (van den Akker 2006; van Tongeren 2004, 2007; Walter 1996). What is the impact of this kind of 'presence ritual' on the survivors? What is the prevailing effect? Are they consoled by the memories and associations evoked by the music, or does the music rather confront the surviving relatives with the harsh reality of the absence of the dead person, although it seems to invite them to an encounter with the loved one? In his book *A Brief History of Death*, the anthropologist and theologian Douglas Davies argues that architecture, literature and music have always spoken 'words against death' (Davies 2002; 2005b: 110). The sounding memory of the life of the deceased is 'sound against death'. The romantic sounds of the repertory of cremation rituals emphasise this. They represent a softening of the harsh reality of death: comfort music, reassuring sounds, sounds that counter death.

Consolationscape

In the title of my chapter, I refer to the crematorium as a 'consolationscape'. The suffix '-scape' is reminiscent of 'landscape', an ever-changing vista that can never be captured in one notion. The Indian anthropologist Arjun Appadurai has used the suffix to define five -scapes as cultural dimensions of globalisation that characterise the current world: ethno-scapes, media-scapes, techno-scapes, finan-scapes, and ideo-scapes (Appadurai 1996). None of these aspects of our current world can be captured by a single notion. They all refer to complex, layered and differentiated phenomena. The same can be said of con- solationscapes. On the one hand, this concept refers to spaces and places where the bereaved find comfort and consolation. On the other hand, it refers to the very personal interpretation of what and who is capable of offering consolation. Transformed into a ritual and musical place, the crematorium might be a place of consolation for the bereaved. Consolation as an abstract, hard-to-theorise concept, might be found in a piece of music, a symbol, or a poem (Barnard 2013). Music and ritual might turn the crematorium into a consolationscape, but that notion is fluid and subject to contesting interpretations. In my con- tribution to this book, I have analysed how the experience of music can be a consoling experience. But the concept of a 'consolationscape' makes me realise that music as part of cremation rituals might not be consoling at all.

References

van den Akker, P. 2006. *De dode nabij: Nieuwe rituelen na overlijden.* Tilburg: IVA.

Ansdell, G. 2014. *How Music Helps in Music Therapy and Everyday Life.* Farnham: Ashgate.

Appadurai, A. 1996. *Modernity at large: Cultural dimensions of globalization.* Minneapolis, MN: University of Minnesota Press.

Barnard, M. 2013. Klanglandschaften des Heiligen jenseits der Hymnologie: Liturgisches Ritual in multikulturellen Kontexten. *IAH Bulletin,* 41, 21–33.

Bell, C. 1997. *Ritual: Perspectives and Dimensions.* New York: Oxford University Press.

Bryant, C. D., ed. 2003. *Handbook of Death & Dying.* Thousand Oaks, CA: Sage.

Cramwinckel, L. 2011. Metamorfose in crematoriumarchitectuur: terreinverkenning van recente ontwikkelingen op het gebied van vormgeving van een nieuw algemeen basaal-sacrale ruimte. Unpublished MA Thesis, Tilburg University.

Davies, D. J. 2002. *Death, Ritual, and Belief: The Rhetoric of Funerary Rites.* London: Continuum.

Davies, D. J. 2005a. *Encyclopedia of Cremation.* Aldershot: Ashgate.

Davies, D. J. 2005b. *A Brief History of Death.* Malden, MA: Blackwell.

Dodge, M. and Kitchin, R. 2005. Code and the transduction of space. *Annals of the Association of American Geographers* 95/1, 162–180.

Feiereisen, F. and Merley Hill, A. 2012. *Germany in the Loud Twentieth Century: An Introduction.* New York: Oxford University Press.

Grainger, H. 2005. *Death Redesigned: British Crematoria, History, Architecture and Landscape.* Reading: Spire Books.

Grimes, R. L. 1995. *Marrying & Burying: Rites of Passage in a Man's Life.* Boulder, CO: Westview Press.

Grimes, R. L. 2014. *The Craft of Ritual Studies.* Oxford; New York: Oxford University Press.

Harmon, K. 2008. *The Mystery We Celebrate, the Song We Sing: A Theology of Liturgical Music.* Collegeville, MN: Liturgical Press.

Honing, H. 2012. *Iedereen is muzikaal: Wat we weten over het luisteren naar muziek.* Fifth, revised edition. Amsterdam: Nieuw Amsterdam.

Hoondert, M. J. M. 2012. Contemplatieve abdijen als musical spaces: vervreemdend en aantrekkelijk. *Jaarboek voor Liturgie-Onderzoek / Yearbook for Ritual and Liturgical Studies* 28, 51–64.

Hoondert, M. J. M. 2014. Muziek, rouw en troost: rouw vanuit muzikaal perspectief, in *Handboek Rouw, Rouwbegeleiding, Rouwtherapie: Tussen presentie en interventie,* edited by J. Maes and H. Modderman. Antwerpen: Witsand Uitgevers, 396–406.

Hoondert, M. J. M. 2015. Het crematorium: ruimte voor rituelen en rouw: Beleid van Crematorium Tilburg ten aanzien van rituelen, herdenkingen en rouw. *Tijdschrift voor Religie, Recht en Beleid* 6/1, 82–94.

Hoondert, M. J. M. and Bruin-Mollenhorst, J. 2016. Music as a lens to study death rituals. *Jaarboek voor Liturgie-Onderzoek / Yearbook for Ritual and Liturgical Studies* 32, 87–104.

Johnson-Laird, P. N. and Oatley, K. 2008. Emotions, music, and literature. *Handbook of Emotions,* edited by L. Feldman Barrett, J. M. Haviland-Jones, and M. Lewis. New York: Guilford Press, 102–113.

Kivy, P. 1990. *Music Alone: Philosophical Reflections on the Purely Musical Experience.* Ithaca, NY: Cornell University Press.

Klaassens, M. and Groote, P. 2012. Designing a place for goodbye: the architecture of crematoria in the Netherlands. *Emotion, Identity and Death: Mortality across Disciplines*, edited by D. J. Davies and C.-W. Park. Farnham: Ashgate, 145–159.

Klaassens, M. and Groote, P. 2014. Postmodern crematoria in the Netherlands: a search for a final sense of place. *Mortality* 19/1, 1–21.

Koopman, C. 1999. Muzikale betekenis in veelvoud: een poging tot ordening. *Tijdschrift voor Muziektheorie* 4/2, 24–33.

Koopman, C. and Davies, S. 2001. Musical meaning in a broader perspective. *Journal of Aesthetics and Art Criticism* 59/3, 261–273.

Leeuw, K. de. 2009. *Leven met de dood: 25 jaar Crematorium voor Tilburg en Omstreken*. Tilburg: Drukkerij Gianotten.

Lefebvre, H. 2004. *The Production of Space*. Malden, MA: Blackwell.

Lukken, G. 2005. *Rituals in Abundance: Critical Reflections on the Place, Form, and Identity of Christian Ritual in Our Culture*. Leuven; Dudley, MA: Peeters.

Maddrell, A. 2013. Living with the deceased: absence, presence and absence-presence. *Cultural Geographies* 20/4, 501–522.

Maddrell, A. 2016. Mapping grief: conceptual framework for understanding the spatial dimensions of bereavement, mourning and remembrance. *Social & Cultural Geography* 17/2, 166–188.

Mollenhorst, J., Hoondert, M., and Zaanen, M. van. 2016. Musical parameters in the playlist of a Dutch crematorium. *Mortality* 21/4, 322–339.

Notermans, C. and Kommers, H. 2012. Researching religion: the iconographic elicitation method. *Qualitative Research* (published online 25 September 2012).

Nugteren, A. 2013. Sensing the 'sacred'? Body, senses and intersensoriality in the academic study of ritual. *Jaarboek voor Liturgie-Onderzoek/ Yearbook for Liturgical and Ritual Studies* 29, 49–65.

Parsons, B. 2012. Identifying key changes: the progress of cremation and its influence on music at funerals in England, 1874–2010. *Mortality* 17/2, 130–144.

Rickard, N. 2004. Intense emotional responses to music: a test of the physiological arousal hypothesis. *Psychology of Music* 34/4, 371–388.

Robinson, J. 2005. *Deeper than reason: Emotion and Its Role in Literature, Music, and Art*. Oxford: Clarendon Press.

Scruton, R. 1999. *The Aesthetics of Music*. Oxford: Oxford University Press.

Small, C. 1998. *Musicking: The Meanings of Performing and Listening*. Middletown, CT: Wesleyan University Press.

Stein, A. 2004. Music, mourning, and consolation. *Journal of the American Psychoanalytic Association* 52/3, 783–811.

Van Tongeren, L. 2004. Individualizing ritual: the personal dimension in funeral liturgy. *Worship* 78, 117–138.

Van Tongeren, L. 2007. *Vaarwel: Verschuivingen in vormgeving en duiding van uitvaartrituelen*. Kampen: Gooi en Sticht.

Walter, T. 1996. A new model of grief: bereavement and biography. *Mortality* 1/1, 7–25.

5 Emotional landscapes

Battlefield memorials to seventeenth-century Civil War conflicts in England and Scotland[1]

Dolly MacKinnon

Introduction

In the twenty-first century memorials to religious wars that physically and emotionally scarred combatants and non-combatants alike during the seventeenth century, can be found in English and Scottish landscapes in unequal ways. This chapter analyses two examples of post-Reformation battlefield memorials to the seventeenth-century civil wars in England, and the Killing Times (c.1685) in Scotland, demonstrating the inclusions and exclusions of the inscriptions incised on those commemorative memorial stones erected. I focus on two Protestant sites – a battlefield at Naseby (1645) in Northamptonshire, and a site of summary execution at Wigtown (1685) in Dumfriesshire Galloway – that have a long history of memorialisation, commemoration, and pilgrimage spanning four centuries. This chapter demonstrates the ways in which these sites form an integral part of each community's invisible consolation landscape in the wake of such conflicts. It charts collective memorialisation as well as 'individual mappings of bereaved people's experiences of significant spaces/places' (Maddrell 2010: 123). These processes of consolation and memorialisation in both a collective and individual sense provide 'emotional maps [that] impact on particular places' and spaces over time (Maddrell 2010: 123). This chapter also addresses the gendered realities of memory and landscapes as sites of consolation, and the practices of histories and history writing, when compared with the material culture of the stone monuments erected over time.

These seventeenth-century sites, and the life and death events that took place there, are still commemorated today, and are sites of pilgrimage, consolation, and contestation. These death sites 'are both intensely private and

1   This research was funded by the Australian Research Council Centre for Excellence in the History of Emotions, Europe 1100–1800 (CE110001011) through my Associate Investigator Grant (2011–2014) 'Emotional Landscapes: English and Scottish Battlefield memorials 1638–1936', and my collaborative Australian Research Council Discovery Grant DP:140101177: 'Battlefields of memory: places of war and remembrance in medieval, and early modern England and Scotland' (2014–2016). The writing was done as part of my Visiting Fellowship at the Institute of Advanced Studies in the Humanities at the University of Edinburgh in early 2016.

personal' while 'simultaneously' providing 'experiences' that are 'expressed often collectively and publicly' (Maddrell and Sidaway 2010: 2). As Christoph Jedan and Erix Venbrux have conceptualised for consolation landscapes, there are 'manifold ways in which human beings respond to bereavement' (Jedan and Venbrux 2013). In keeping with this, I also draw upon the influential work of Avril Maddrell that analyses 'living with the deceased' in landscape settings (Maddrell 2013a: 501–522). As Maddrell expressed, from self-reflexion surrounding her own personal experiences of death, 'grief and mourning were framed within a detailed topography of significant spaces and practices' (2016: 168). This chapter focuses on the physical monuments (rather than the printed memorialisation, and other forms of performance that surround these sites) in these significant battlefield and summary executions spaces that express, to the following generations, a sense of consolation. For Naseby, there was a trust in hope over fear that obedience between subjects and the monarch would be perpetuated, while for Wigtown there was a belief in an ongoing Reformation that ensured liberties and freedoms for future generations through the death of martyrs. These ongoing struggles spilt the blood of combatants (soldiers), non-combatants (men, women and children) of the baggage trains that accompanied opposing armies, as well as civilians (women, men and children), caught up in the conflict. Variously described by contemporaries as martyrs, rebels, or the winners – losing and winning sides were loyal to their own version of the true religion. Similar to notions of worlds within worlds, these simultaneous emotional communities committed these events to memory, for both the winning as well as losing sides reiterate their significance through a reciprocal process of pilgrimage, remembering and consolation over time.

Emotional communities

In using the term 'emotional communities' I am referring to and adapting Barbara Rosenwein's conceptualisation (Rosenwein 2006 and 2015: 249). Rosenwein contends that for emotional communities to communicate within and between each other they must have a shared emotional language that was readily understood (Rosenwein 2006 and 2015: 249). Therefore, it is through these texts on the commemorative monuments to seventeenth-century conflicts that we can see and read the emotional styles of the shared community within the landscape at or near the specific places of death. If we understand emotional communities to exist, then how do these emotional communities create their emotional landscapes, and what traces of those early modern landscapes are still evident for Naseby (1645) and Wigtown (1685) in the intervening centuries up until the present? What trans-generational uses and traces, since the seventeenth century, are overlayed in this heterotopic and polysemic landscape? Furthermore, why do these emotional narratives come into existence, and what role do they play in consolation landscapes?

From a methodological standpoint this chapter takes as its starting point the innovative work of Monique Scheer (2012: 193) who articulated the potential of 'emotional practice' as a framework for analysing and understanding communities using their emotions through actions. As I will demonstrate, emotions are not simply internal entities. Citing the philosophical work of Robert C. Solomon (2007: 155, 157), Scheer (2012: 194) challenges us to consider 'emotions as acts', realising that emotions are 'not entities *in* consciousness' but rather 'acts of consciousness'. Emotions become not something we 'just have', but rather they 'are indeed something we do' (Scheer 2012: 194). As Scheer notes, 'emotional practices can be carried out alone, but they are frequently embedded in social settings' (2012: 211). In order for emotional practice to function within emotional communities (if you like, emotional worlds within worlds) then each dynamic emotional community must be actively 'mobilizing, naming, communicating and regulating emotions' (Scheer 2012: 194, 209–220). Emotional practices 'are stored in the habitus, which provides socially anchored responses to others'; 'emotion-as-practice is dependent and intertwined' with 'speaking, gesturing, remembering, manipulating objects, and perceiving sounds, smells, and spaces' (Scheer 2012: 209, 211). All of these elements are dynamic in emotional ritual practices 'as a means of achieving, training, articulating, and modulating [the] emotional for personal as well as social purposes' (Scheer 2012: 210). As I will demonstrate in this chapter, emotional practice is a dynamic process over time, and also within shorter and longer timeframes. Scheer concludes that 'centuries of reflection on the effects of observing others' bodies and voices on the stage, on the soapbox, or in the pulpit have elaborated, refined, and revised emotional practices' (Scheer 2012: 211). For the purposes of my analysis of those at Naseby (1645), and the Cameronian community of Covenanters at Wigtown (1685), who followed the preaching of the radical field preacher Robert Cameron (1648–1680), I am 'thinking hard about what people are *doing*' in Scheer's sense (2012: 217). It is 'the specific situatedness of these doings' that I am analysing by 'trying to get a look at bodies and artefacts of the past' in their specific sites of consolation (Scheer 2012: 217).

Naseby (1645)

By sunset on 14 June 1645, as news of the civil war battle's outcome between the parliamentary and royalist forces began to spread across the countryside Naseby-Field, part of a rural farming community in Northamptonshire in England, was transformed into a Parliamentary victory in the national mnemonic. Naseby had confirmed its place as a pivotal battle in these seventeenth-century religious wars. The decisive rout of King Charles I, Prince Rupert and the Royalist forces by Sir Thomas Fairfax, Oliver Cromwell et al. fighting in the Parliamentary army earlier that day, had resulted in over one thousand Royalist and one hundred and fifty Parliamentarian casualties, as well as the slaughter of over one hundred women in the Royalist baggage train. The turf of Naseby battlefield was soaked with the split blood of liberty and loyalties, with the bodies of men and women forming the tangible evidence of an irrevocable political and religious divide. Naseby was a turning point in this conflict, but one in which the features of gendered memorialisation were to be enshrined (Stoyle 2008: 895–923). Those men and women who died there were buried in large pits, while the walking wounded (those men and women who left the field), were marked with lifelong emotional and physical battle scars. Some of those who left the field alive would later die of their wounds (Forde 1995: 302). Many of the dead would be buried in pits, the locations of which would over time be forgotten. For example, in May 1702 during a process of digging for gravel a burial pit was rediscovered (Vialls and Collins 2004: 172). One eye witness to the aftermath of the Naseby battle in 1645 claimed 'I saw the field so bestrewed with Carcases of Horse and Men, as was *most sad to behold*, because Subjects under one government' and yet the emotion was clear 'but most happy in this, because there were most professed enemies of God' (Gentleman in Northampton, 1645: 3–5). What is more the word 'Naseby' became a short hand synonymous with this emotional nature and consequences of this battlefield. Elizabeth Isham (1608–1654) wrote in her diary as she passed through the county of Northamptonshire, just the name 'Naseby Field'. It was only later that Isham added 'God be praised at [Charles I] scaping' (MacKinnon 2015: 218). Isham added the emotional context to her diary entry, which in Scheer's approach, demonstrates the ways in which 'emotions are perhaps most obviously practiced when they are involved in communication' (2012: 214). Naseby is the point of consolation in the landscape and it is precisely 'because people know that emotions do things in social contents', that Isham uses 'God be praised at [Charles I] scaping' [sic] in Scheer's sense 'as means of exchange' (Scheer 2012: 214).

Five monuments commemorating the men of the battle were erected at Naseby from the eighteenth century to the early twenty-first century: a monument was erected by George Ashby (1725–1802) in a private rustic grove in 1771 at Haselbeech looking towards the Naseby parish church; in the late eighteenth century the Right Hon. Lord Viscount Cullen erected a monument in his private garden in sight of the battlefield of Naseby Field from Rushton Hall (1780s); the Fitzgerald monument was erected on a parcel of land at the highest point in Naseby, and indeed Northamptonshire, to the west of the actual battlefield (1823); a monument in 1936 claiming it was 'correcting an error in history' was erected by the Cromwell Association on the site of the battlefield; and in the twenty-first century a monument to a Royalist Regiment was erected at Naseby. Each monument is an emotional and cultural expression in and over time that 'is unique and gets put into a category – is typified – only through naming' making emotion and consolation tangible (Scheer 2012: 213).

Wigtown (1685)

Forty years after the battle of Naseby (1645) in England, to the north over the English border into Dumfriesshire Galloway in Scotland on 11 and 12 May 1685 during the height of the Killing Times, the civil wars of religion continued with great intensity. Five Covenanters (two women and the next day three men), adherents of the Cameronian society, were executed at Wigtown. Two women were drowned and three men were hanged. In the twenty-first century the Wigtown landscape is marked by a heritage trail to the martyrs of the Killing Times, or alternately Wigtown is described as a book town. Like Naseby, gender colours the inclusions and exclusions of conflict and the memorials erected to commemorate those involved in these conflicts. The existence of the female Wigtown martyrs has been commemorated, contested and by some categorically denied. Here there are three Wigtown martyrs' monuments: individual monuments (two women and three men) in the Kirkyard erected in *c.*1714, some 30 years after the events; a collective memorial to both the women and the men that was erected by public subscription in 1858; and the *c.*1937/8 stone stake for the female martyrs replacing an earlier wooden stake said to mark the site of the drownings in 1685. This traditional practice of marking death in the landscape was a shared common response to sudden death (Maddrell 2016). For example, in Scotland the Reverend Mr Robert Laws recounted the case of a 'dragoon [who accidentally shot himself] ... through the heart' after he had stopped at the roadside 'to ease nature' (1818: 230). Laws added that 'he is buried where he died, and a small heap of stones cast upon him, as a remembrance of that fact' (1818: 230). Here death is given a material marker through the placement of stones of remembrance which were intended to commit the event to memory, and to mark this lesson in the emotional landscape of the community as a site of consolation and pilgrimage (MacKinnon 2019: 163–178). There was no need for a formal inscription, as word of mouth within this emotional community conveyed the nature of these tragic events overtime. The stone, or wooden stake, in the case of the Wigtown women, claimed to be the location of the summary execution.

The Covenanter monuments (erected by Calvinist Presbyterians) to the Wigtown martyrs represent part of the Covenanting community's emotional inscription into that landscape, first through memory and then through remembrance, of a violent and bloody period in Scottish history. The Covenanters were the adherents of the National Covenant (1638), Solemn League and Covenant (1641/2), and found their loyalty challenged and their faith and religious worship tested by the Crown and the state kirk. They were driven out of their kirks to worship in clandestine conventicles held in the fields and on the moors until 1685. Either preaching at or attending, what were described as fanatical houses or field meetings, was outlawed, and punishable by death and the confiscation of goods (Act against preachers at conventicles and hearers at field conventicles, 28 April

1685). Each martyr's individual adherence to King Jesus, rather than the Stuart Crown and monarchy, was demonstrated by their refusal to take the Oath of Abjuration (that is to swear loyalty to the crown over the Covenants, and to refuse to fight in open rebellion) during the 'Killing Times' of 1680–1685. In the centuries that followed the Wigtown martyrs' monuments and their pilgrimage trails, demonstrate how, over time, the memory of these events was transformed into communal expressions of remembrance, by actively re-inscribing this emotional landscape after the initial violent events. The nineteenth century saw the rehabilitation of the Covenanters for the purposes of the creation of Scottish identity. James J. Coleman has discussed how 'different localities within Scotland invoked Covenanting memory as a means of celebrating their own contribution to Scotland's nationality' from the nineteenth century onwards for the express purposes of defining Scottish identity and nationalism in the present (Coleman 2014: 135–153).

The emotional topography of both the Naseby (1645) and the Wigtown (1685) Covenanter memorials demonstrates the powerful emotions of silence, memory, remembrance and amnesia. The monuments, and their crafted inscriptions, offer consolation to those who resided in or made a pilgrimage to these landscapes. The pivotal religious events were perpetuated first in popular memory, and then eventually through physical monuments and inscriptions, moulded into versions of national and religious history in the English and Scottish landscapes. What emerged was the gendered nature of those memorialisations. The inscriptions demonstrate clear distinctions between perceptions of male battlefields at Naseby (1645), and the memorialisation of individual male and female martyrs at Wigtown (1685). In addition, these monuments make crucial distinctions between combatants and non-combatants (women, men and children), and the place or absence of women in their historical narratives. Here is the tradition of historical amnesia in the construction of national histories regarding the true human costs of war. Emotional and traumatised communities, such as the Covenanter community in and around Wigtown, could and did erect monuments that were at odds with the national historical polemic. Victorious histories were written from Royalist and an Episcopalian perspective (that is, church government by bishops, which the Covenanters vehemently opposed). These histories not only challenged the importance of the alleged Covenanter martyrdom, but even denied the existence of the Wigtown women, though never questioning the existence of the Wigtown men. The actions of the Covenanters were reduced to a historiographical abbreviation summed up as 'the martyrology and mythology of the "Killing Times"' that 'has often dominated the historiography of this entire period despite the fact that it was a short-lived aberration affecting only the adherents of one tiny Presbyterian faction' (Lynch 2001: 114). What Michael Lynch fails to recognise is that 'these "almost 100 individuals" were connected to their wider families, as well as their religious communities, resulting in the numbers of those effected by these violent events' rising 'into the hundreds' (MacKinnon, 2019: 170). As I have shown elsewhere,

> what the political historians have categorised as isolated and individual acts of violence actually generated an emotional ripple and resonance that travelled well beyond those present at such killings, and has fueled the centuries following. Each individual was a family member, a sibling, spouse, aunt, or uncle, grandparent, a cousin, or a spiritual friend.
>
> (MacKinnon 2019: 170)

At Naseby, the men on both sides of the conflict (parliamentarians and royalists) fought and died, and their presence in this landscape was never questioned. It was the presence and deaths of women in the Royalist baggage train that was challenged.

Emotional geographies

As Kay Anderson and Susan J. Smith have concluded, 'thinking emotionally is implicitly cast as a sense of subjectivity which clouds vision and impairs judgment, while good scholarship depends on keeping one's own emotions under control and others' under wraps' (Anderson and Smith 2001: 7–10). Yet the local communities connected with battlefield trauma and death form part of what Anderson and Smith would call 'the human world' as 'constructed and lived through emotions' (Anderson and Smith 2001: 7). These emotional communities can create a process of memorialisation and commemoration, and ultimately an emotional landscape in which to give voice to their particular views even when they are at odds with state histories. These landscapes also offer places of consolation and commemoration. Here monuments act as tangible forms of competing consolations – individual/familial memories and stories versus national histories. In the case of the Wigtown Martyrs, sections of the community preserved the memories, ensured remembrances and erected memorials that articulated their presence in physical form (stone monuments) from c.1714 onwards. These monuments offered consolation to those left behind who witnessed the deaths of Protestant martyrs and who lived on, during a period of fear, which the adherents of the Covenant understood to be a process of ongoing Reformation where the outcome was yet to be determined. To cite C. J. Withers, from another context, 'memory may be strongly embodied and situated in the landscape…but it is not just placed in any simple geographical sense of that term … [rather] it resides in texts, practices and in the struggle over its meaning' (Withers 2004: 316–339, cited in Maddrell 2013b: especially 687). For example, death stones were a common practice in Scotland, where the site of a death was marked by the placement of a stone, or stones in the form of a carne. Formal monuments with inscriptions could be erected later, as was the case at Wigtown, once the political and religious environment allowed such displays to exist. The executions at Wigtown had occurred in 1685, but it was not until 1704 that formal discussions in the Kirk at Wigtown led by the new minister Mr. Thomas Kerr, openly acknowledged the past event. One man, Bailie M'Keand, who had been involved came to the kirk with the matters weighing heavily upon his conscience, and 'declaring his grief of his heart that he should have sitten in the seize of these women who were sentenced to die in this place in 1685' (Minutes, Presbytery of Wigtown, 8 July 1704). The Presbytery 'being satisfied with his conversation since, and the present evidence of repentance now, they granted him privilege' once more to take the sacrament (Stewart 1869: 95).

The competing debates surrounding the fact or fiction of the death of the female martyrs at Wigtown, as well as the deaths of the women of the Naseby Baggage train, form part of this ongoing struggle for stable meanings in the landscape. As one nineteenth-century commentator observed about the Kirkyard monuments to the Wigtown martyrs, 'Is it possible then to conceive any one venturing to commit such an outrage on truth and propriety, as to inscribe on a tombstone in a churchyard visited every Sunday by the whole population of a country town, what they all would have known to be a "fable"?' (Stewart 1869: 103). Here the physical monument bears witness to the emotional and intangible landscapes of consolation.

Both Naseby and Wigtown are contested sites, where the absence of the women in the baggage train in any formal memorials at Naseby, and the denial of the very existence of the Wigtown female martyrs, even when memorials were erected to them (*c.*1714, 1858, and *c.*1937), points to the gendered notions of heterotopic memorial space in these places, as well as the masculinised understanding of objectivity and battlefields, and the feminisation of subjectivity, emotion and female sites of summary execution. These sites simultaneously represent presence and absence, recall and amnesia, consolation for some, but not for all. For Naseby, the two monuments erected in 1824 and 1936 actively dampened down the emotional reality of civil war, opting in favour to cast the events as a valuable and consoling lesson well learnt, or simply establishing the importance of battlefield facts in order to cancel out existing errors. The Wigtown monuments play to the emotion of the events of the past as part of a lineage in the history of the struggle for religious freedoms and liberties enjoyed in the centuries to follow. These religious liberties are attributed to the consolation and sacrifice of the seventeenth-century covenanter martyrs. Memorialisation, including those at Wigtown, creates a virtual community in a heterotopic space.

The Naseby monuments

Central to the process of commemoration and consolation is access to the landscape for the purposes of remembrance. The first monument erected to the Naseby battle of 1645 was placed not on the battlefield itself, but rather on private land in a Rustic Altar in a Grove at Haselbeech, Northamptonshire, 'in sight of Naseby Church', and dated 30 January 1771. It was erected by George Ashby (1725–1802) who leased Haselbeech Hall (MacKinnon 2015: 219). 'The Ashby family was descended from the royalist family connected with Ashby de la Zouch Castle in Leicestershire' which was 'held for the king during the English Civil War' (MacKinnon 2015: 219). The 1771 monument's inscription reads:

> SACRED TO MONARCHY, FREEDOM, AND PEACE, THIS SMALL MONUMENT WAS PLACED, IN SIGHT OF NASEBY CHURCH, 30ᵗ JAN. 1771, BY G. ASHBY. MAY THE BEST OF KINGS BE AFRAID OF NOTHING SO MUCH AS INCROACHING UPON THE RIGHTS OF THE PEOPLE! MAY THE SUBJECT, HOWEVER FOND OF LIBERTY, BE QUIET, BE THANKFUL, SO LONG AS HE HAS NO GRIEVANCES TO BE REDRESSED!
>
> (Mastin, 1792: 200)

The antiquarian John Mastin transcribed the text that 'was written for a Rustic Altar in a Grove at Haselbeech, Northamptonshire' (1792: 200).

The significance of this parable lies in the date of commemoration, and the claims for freedom and peace. The date, 122 years earlier, commemorated the execution, of what many seventeenth-century contemporaries considered to be, Charles Stuart, that 'man of blood' [Charles I], outside White Hall in the depth of winter: 30 January 1649. Naseby (1645) formed a crucial part in the inevitable sequence of events in the national story of remembrance and consolation. But this monument did not just speak to the past, it also heralded a warning to future communities of the perils of conflict and the folly of civil war. The 1771 monument predates the America Revolution (1775–1783), acting as a mechanism through which the past offers a parable of temperance and consolation through a process of commemorating a past designed to instruct future generations about the turmoil of civil wars sparked by internal state disputes between subjects and their sovereign.

The second monument at Naseby was erected by the Right Hon. Lord Viscount Cullen, and was placed in a private 'alcove commending a view of Naseby Field' at Rushton Hall in Northampton. The inscription was written by 'Dr. Bennet, Bishop of Cork' (Lockinge 1830: 122–123), and read:

> Where yon blue field scarce meets our streaming eyes,
> A fatal name for England, NASEBY [1645] lies.
> There hapless Charles beheld his fortune cross'd
> His forces vanquish'd, and his kingdom lost.
> There gallant Lisle [Sir George Lisle 1610–1648] a mark for thousands stood,

And Dormer [Sir Robert Dormer of Wing, 2nd Baronet, 1st Earl of
Carnarvon, Viscount Ascott, 1610–1643] seal'd his loyalty in blood;
Whilst down yon hill's steep side with headlong force,
Victorious Cromwell chac'd the northern horse.
Hence Anarchy our church and state profan'd,
And tyrants in the mask of Freedom reign'd,
In times like these, when party bears command,
And faction scatters discord thro' the land;
Let these sad scenes a useful lesson yield,
Lest future NASEBYS rise in every field.

(Lockinge 1830: 122–123)

The 'tablet' was 'in an alcove in a wilderness, planted, with hornbeam
hedges, in the form of a great cross' (Neale 1826: Vol. III, 1–6, 'Rushton Hall,
Northamptonshire; the seat of the Hon. Mrs Cockayne Medlycott'). Here the
inscription conflates and expands time by including events both before and
after the battle of Naseby in 1645. For example, the death of Sir Robert
Doermer of Wing, 2nd Baronet, 1st Earl of Carnarvon, Viscount Ascot (1610–
1643), occurred at the Battle of Newbury on 20 September 1643. Sir George
Lisle (1610–1648), while wounded at the Battle of Naseby, and knighted
shortly afterwards, was later executed in 1648, along with Sir Charles Lucas
after the Siege of Colchester in Essex. This Naseby monument warned future
generations of the perils of civil war conflict that might 'rise in every field' if
left unchecked (MacKinnon 2015: 221). Here the monuments offered consolation only to those with access to this privileged social space who were
families with royalist leanings.

In 1823 a third monument at Naseby, an obelisk, was erected at the highest
geographical point both in Naseby, and Northamptonshire, with the battlefield situated far to the west of this site, on the private land of John and Mary
Fitzgerald (MacKinnon 2015: 223). The Fitzgeralds' inscription stated that
the monument was intended:

To commemorate that great and decisive battle fought in this field on the
XIV day of June MDCXLV [1645] between the Royalist Army commanded
by His Majesty King Charles the first, and the Parliament Forces headed by
the Generals Fairfax and Cromwell, which terminated fatally for the Royal
cause, let to the subversion of the throne, the altar, and the constitution,
and for years plunged this nation into the horrors of anarchy and civil
war: leaving a useful lesson to British Kings never to exceed the bounds
of their just prerogative, and to British subjects never to swerve from the
allegiance due to their legitimate monarch. This Pillar was erected by
John and Mary Francis Fitzgerald, Lord and Lady of the Manor of
Naseby:

A.D. MDCCCXXIII [1823]

Figure 5.1 Naseby Memorial, 1823

As I have demonstrated elsewhere, it was the ownership of private land that enabled any memorial practices, driven by the owner, to either take place or be blocked (MacKinnon 2015: 225). It was not until 1936 that a monument was actually erected on the battlefield site itself by the Cromwell Association, at the behest of Mr. C. E. Reich, and designed by architect J. A. Gotch. The intent was to 'correct an error in history' and Reich hoped 'you and your children, and your children's children will remember that here at Naseby occurred one of the most important events in the history of your country' (MacKinnon 2015: 226–227).

In 2012, an historical re-enactment society erected a monument to a Royalist Regiment at Naseby (MacKinnon 2015: 229). The collective and national appropriation of this emotional landscape continues to be contested. The gendered memorial practices also continue, with the women of the Battle

Figure 5.2 Naseby Memorial, 1936

of Naseby unacknowledged in formal monuments (though included in late twentieth-century interpretation boards) and the ongoing questioning of their existence and numbers are still debated in the historiography. Naseby is a heterotopic landscape of commemoration and amnesia.

Wigtown monuments

In *c.*1714 the first individual funeral monuments were erected in the Wigtown Kirkyard, some 30 years or so after the events at the height of the Killing Times in 1685. These monuments then mark publicly in the Kirk Yard at Wigtown, the burial places that had until that time, through fear, only been marked by the cut in the ground and by individual and community memory. The monuments name those who died over a two-day period at Wigtown: Margaret Willson, and Margaret Lachlane who were drowned on 11 May 1685, and William Johnston, George Walker, and John Milroy who were hanged on 12 May 1685. Those who had witnessed these events had also committed to memory the site of the burials of the women's bodies in the Kirkyard after they were reclaimed at the low tide in 1685. No mention is made of the reclaiming of the men's bodies from their site of execution. By the early eighteenth century, Margaret Willson's brother Thomas was said to have returned to the Wigtown area where he remained until 1734. It is thought that Thomas had the monuments erected in the kirk with community support. The monuments' inscriptions rehearse the earlier accounts that had appeared in print about these events as the stories of the martyrdom (first as images only and then as named individuals) proliferated from the presses from 1687 onwards (Shields, 1687, Frontispiece image). The individual inscriptions are emblematic of an ongoing religious struggle in which the women and men, who died for King Jesus, were considered individual Christian soldiers, martyrs, in God's Reformation army.

> [At right angles to the main inscription, at side of stone]
> Here lies Margret
> Willson Doughter
> to Gilbert Willson
> in Glenvernoch
> who was drouned
> ANNO 1685 Ages 18 C[sic]
> [Main inscription]
> Let earth and stone still witnes bear
> Heir lyes a virgine martyre here
> Murther'd for ouning Christ Supreame
> Head of his Church and no more crime
> But not abjuring presbytry
> And her not ouning prelacy
> They her condem'd, but unjust law,
> Of Heaven nor hell they stood no aw
> Within the sea tyd to a stake
> She suffered for Christ Jesus sake
> The actors of this cruel crime
> Was Lagg.. Strachan. Winram. and Grhame [sic].
> Neither young yeares, not yet old age
> Could stop the fury of there rage.
>   (Campbell 1996: 184–185)

The language of emotions runs through these inscriptions which, in Scheer's sense, by 'putting a name on' these 'feelings is part and parcel of experiencing them' (2012: 209, 212). Readers are drawn into the experiences of death and emotions by reading the inscription.

The monument to Margaret Lachlane reads as follows, and also uses the same type of emotional language:

Me mento mori
Here lies
Margaret Lachlane
Who was by un[-]
just law sentec[-]
ed To Dye by Lagg,
Strachane Win[-]
rame and Grame
and tyed to a
Stake within the
Flood for her
Adherence
To Scotland's Re[-]
formation Cove[-]
nants National
and Solemn League
Aged 63 1685.
[Vertically, at right angles to the ends of the lines, the words]
Surnamed Grier.

(Campbell 1996: 185)

The purpose and function of the inscriptions recounts both the emotional circumstances of their deaths as religious martyrs, and also links them to Scotland's ongoing fight for religious liberties and freedoms into the future. The monument to Johnston, Milroy and Walker also rehearses this engagement with its readers stating:

Mememto mori
Here lyes William Johnston
John Milroy[,] George Walker
Who was without sente[-]
nce of law hanged by Ma[-]
jor Winram for their Adher[-]
ance to Scotland's refor[-
mation covenants natio[-]
nal and solam leagwe.
1685.
    (Campbell 1996: 185–186)

The Wigtown monuments to the events of 1685 represent a site of pilgrimage and a location of community consolation for many but not for all in the Kirkyard at Wigtown. Over a century later in 1858 a second monument was erected in Windy Hill overlooking the town and bay. This 1858 monument commemorated the past in order to justify the events of the Disruption in the Church of Scotland that had occurred in 1848, which resulted in the Free Church of Scotland withdrawing from the established church of Scotland.

The text, therefore, was both a lesson, and a form of consolation, for the inscription reads:

This monument has been erected
In memory of the noble army of Martyrs of Galloway
And other parts of Scotland by whom
During the age of persecution our
Religion and liberties now established
Were secured: and As a lesson
To their posterity never to loose or abuse
These glorious principle, planted by their
Labours rooted in their sufferings
And watered with their blood.

The Wigtown martyrs monument of 1858 commemorates the women and the men, and also speaks in sympathy with the earlier individual monuments erected in the Kirkyard by 1714. The consolation narrative for this landscape is strengthened through this process of memorials over time. These sites are highly emotional ones that have become sites of pilgrimage and consolation for those who have visited them over the following centuries. They are simultaneously viewed as martyrs by religious pilgrims in the twentieth century and just fictional stories by those who do not believe in their historical existence.

In 1937/8 a stone monument at Wigtown replaced an earlier wooden stake that was said to have stood on the very place where Margaret Willson and Margate McLachlane had been drowned, marking the third site of memorialisation and consolation. Whether this is the exact site of the martyrdom is unclear, as the river channels are constantly shifting and changing as the coastlines erode and accrete through shifting sands and sediment. Yet here the perceived site of the drownings was also commemorated. The more the existence of the women was contested, the more the landscape was used to place and commemorate the women's presence in the landscape. Wigtown's modern harbour was moved to this current site in 1825, and therefore the conflation of the past and the present is evidence here too. There is no monument commemorating the site of the hangings of Johnston, Milroy and Walker. In the gendered interpretation of the past, the men's historical presence does not need a tangible monument in the present to ensure its validity in the historical narrative. These men may have been hanged in a public space, such as outside the Tollbooth (jail), in the High Street in Wigtown. Landscape, like memory, remembrance and emotion, shifts and changes, losing some elements and gaining others in the ongoing process of landscape re-formation and consolation. Its gendered nature persists.

Conclusion

These monuments erected for Naseby (1645, 1771, 1823, and 1936), and Wigtown (1685, *c.*1714, 1858 and 1937) offer consolation by linking the past with the present and future. The emotional Covenanter monuments exemplify the battle of Christian soldiers in an ongoing Reformation, while the four monuments at Naseby represent a parable of a valuable lesson learnt by King, church, constitution, country, and subject. Furthermore, the Covenanter monuments are individual, as well as collective, representing the women and men involved. For Naseby, the women of the baggage train are missing, and the masculine monuments of Naseby commemorate at an individual level, only a monarch and members of the aristocracy and gentry. The Wigtown women's martyrdom is challenged, and the men's is unchallenged.

Taking Helen Armstrong's discussion, from another context, about the inherent tensions between history, landscape and tourism, the sites of Naseby and Wigtown are tourist destinations. The current proposed £8 million Naseby Interpretation Centre reflects 'the tension between the mythic qualities of landscape and the current pressure to consume landscapes for pleasure through tourism' (Armstrong 2001: 12). As Armstrong cogently argues, 'landscapes are not merely blank slates on which media can inscribe various meanings. The landscape is already a richly inscribed medium through which meanings of place/space are understood' (Armstrong 2001: 13). In this chapter I have shown the seventeenth-century history of Naseby and Wigtown in order to demonstrate how experiences of death and consolation 'make a deep impression of *sense of* self, private and public identity, as well as *sense of* place in the built and natural environment' (Maddrell and Sidaway 2010: 2). Maddrell and Sidaway have demonstrated that 'these experiences of death, dying and mourning are mediated through the intersections of the body, culture, society and [the] state' (2010: 2). Therefore, the monuments and their inscriptions at Naseby and Wigtown remain dynamic seventeenth-century emotional landscapes, where the contested heterotopic spaces of a continually gendered consolation landscape persist not only into our present but into the future.

References

Act against preachers at conventicles and hearers at field conventicle, 28 April 1685. University of St Andrews, Records of the Parliaments of Scotland to 1707. http://www.rps.ac.uk/ 1685/4/28, NAS. PA2/32, f.153 [Accessed 12 November 2017].

Anderson, K. and Smith, S. J. 2001. Editorial: Emotional Geographies. *Transactions of the Institute of Geography* 26/1, 7–10.

Armstrong, H. 2001. Spectacles and Tourism as the Faustian Bargain: Sustaining the Myths of Landscape. *Spectator* 21/1, 12–25.

Campbell, T. 1996. *Standing Witnesses: An Illustrated Guide to the Scottish Covenanters.* Edinburgh: Saltire Society.

Coleman, J.C. 2014. *Remembering the Past in Nineteenth-Century Scotland: Commemoration, Nationality, and Memory.* Edinburgh: Edinburgh University Press.

Foard, G. 1995. *Naseby: The Decisive Campaign.* South Yorkshire: Pen & Sword.

Gentleman in Northampton. 1645. *A More Exact and Perfect Relation of the Great Victory.* London: John Wright.

Jedan, C. and Venbrux, E. [Session Convenors]. 2013. Consolation-scapes: Analysing grief and consolation between space and culture (Session 2), 4th International and Interdisciplinary Conference on Emotional Geographies, 1–3 July 2013, University of Groningen.

Laws, Rev. Mr. R. 1818. *Memorials … 1638 to 1684.* Edinburgh: Archibald & Co.

Lockinge, H. 1830. *Historical Gleanings on the Memorable Field of Naseby.* London: Longman, Rees, Rome, Brown and Green.

Lynch, M. (ed.) 2001. *The Oxford Companion to Scottish History.* Oxford: Oxford University Press.

MacKinnon, D. 2015. Correcting an error in history: battlefield memorials at Marston Moor and Naseby. *Parergon* 32/3, 205–235.

MacKinnon, D. 2019. This humble monument of guiltless blood: The emotional landscape of Covenanter monuments, in *Writing War in Britain and France, 1370–1854: A History of Emotions*, edited by S. Downes, A. Lynch, and K. O'Loughlin. London and New York: Routledge.

Maddrell, A. 2010. Memory, mourning and landscape in the Scottish mountains: discourses of wilderness, gender and entitlement in online and media debates on mountainside memorials, in *Memory, Mourning and Landscape*, edited by E. Anderson, A. Maddrell, K. McLouglin and E. Vincent. Amsterdam: Rodopi, 123–146.

Maddrell, A. 2013a. Living with the deceased: absence, presence and absence-presence. *Cultural Geographies* 20/4, 501–522.

Maddrell, A. 2013b. A place of grief and belief: the Witness Cairn, Isle of Whithorn, Galloway, Scotland. *Social & Cultural Geography* 10/6, 675–693.

Maddrell, A. 2016. Mapping grief: a conceptual framework for understanding the spatial dimensions of bereavement, mourning and remembrance. *Social & Cultural Geography* 17/2, 166–188.

Maddrell, A. and Sidaway, J. D., eds. 2010. *Deathscapes: Spaces for Death, Dying, Mourning and Remembrance.* Farnham: Ashgate.

Mastin, J. 1792. *The History and Antiquity of Naseby in the County of Northampton.* Cambridge: Francis Hodson.

Neale, J. P. 1826. *Views of the Seats of Noblemen and Gentlemen in England, Wales, Scotland and Ireland*, second series. Vol. III. London: Sherwood, Gilbert and Piper.

Rosenwein, B. H. 2006. *Emotional Communities in the Early Middle Ages.* Ithaca, NY: Cornell University Press.

Rosenwein, B. H. 2015. *Generations of Feeling: A History of Emotions, 600–1700.* Cambridge: Cambridge University Press.

Ryden, K. C. 1993. *Mapping the Invisible Landscape: Folklore, Writing, and the Sense of Place.* Iowa: University of Iowa Press.

Scheer, M. 2012. Are emotions a kind of practice (and is that what makes them have a history)? A Bourdieuian approach to understanding emotion. *History and Theory* 51, 193–220.

[Shields, A.] 1687. *A Hind let loose, or An Historical Representation of the Testimonies, of the Church of Scotland, for the Interest of Christ, with the true State thereof in all its Periods: Together With A Vindication of the present Testimonie, against the*

Popish, Prelatical, & Malignant Enemies of that Church ... : Wherein Several Controversies of Greatest Consequence are enquired into, and in some measure cleared; concerning hearing of the Curats, owning of the present Tyrannie, taking of ensnaring Oaths & Bonds, frequenting of field meetings, Defensive Resistence of Tyrannical Violence ... / By a Lover of true Liberty. Edinburgh.

Solomon, R. C. 2007. *True to Our Feelings: What Our Emotions Are Really Telling Us.* Oxford: Oxford University Press.

Stewart, A. 1869. *History Vindicated in the Case of the Wigtown Martyrs.* Second edition. Edinburgh: Edmonston and Douglas.

Stoyle, M. 2008. The Road to Farondon Field: Explaining the Massacre of the Royalist Women at Naseby. *The English Historical Review.* 123/503: 895–923.

Vialls, C. and Collins, K., eds. 2004. *A Georgian Country Parson The Rev. John Mastin of Naseby.* Northampton: Northampton Record Society.

Withers, C. J. 2004. Memory and the History of Geographical Knowledge: The Commemorations of Mungo Park, African Explorer. *Journal of Historical Geography* 30/2, 316–339.

6 Danish churchyards as consolationscapes

Anne Kjærsgaard

Introduction

To many visitors, Danish churchyards appear highly secularised, but in this chapter I will argue that they in fact materialise religious Protestant norms, especially norms about how to find consolation in the face of death.[1] Due to the character of Protestant material culture, however, we tend to overlook the Protestant norms of consolation that these graveyards embody. The appearance of the churchyards thus does not result from a lack of religion, but rather from a particular form of religion with a particular understanding of material culture and consolation. In this chapter, I will describe how this understanding came about at the time of the Reformation and how it was implemented in different ways in churchyards in Lutheran Denmark. I will further show that the advent of cremation and the changes in the structural design of Danish churchyards this brought about – though normally seen as features of secularisation – strengthened Protestant norms. I thus argue that instead of a withdrawal of official religion, the Lutheran Church has actually increased its grip on the graveyard.

Conflicts concerning the appearance of churchyards are regularly understood within a framework of aesthetics but should instead be considered as religiously based. Because the Protestant norms that graveyards embody are often overlooked, it is not fully recognised how they repress non-prescribed forms of religion and with them connected understandings of consolation. This repression can give rise to conflicts over material culture. The Evangelical Lutheran Church has a near-monopoly on burial grounds in Denmark and therefore almost all Danes end up in a Lutheran churchyard,[2] whether

1 I use the words 'churchyard' and 'graveyard' interchangeably for the Danish word *kirkegård*. I avoid using the word 'cemetery' because no equivalent word exists in the Danish language (for a further explanation of this see below, note 2). The terms 'burial grounds' and 'burial place' are used as broader terms.

2 The Lutheran Church possesses a little over 2100 churchyards or about 99% of all burial grounds in Denmark. In a number of the biggest cities the municipality (*kommune*) run the burial grounds, but with only one exception they have all been consecrated. They are therefore termed *kommunale kirkegårde*. Apart from this a few burial grounds belong to other religious groups (Christian, Jewish and Islamic). This explains why there is no close equivalent for the word 'cemetery' in the

they are members of the church or not. It is to be expected that the regulation of material culture and its implicit Protestant norms can make it difficult to find consolation for the 24.1 per cent of the population that are non-members (Kirkestatistik 2017). However, the same can also be the case for members of the Lutheran Church, as I will show in the last part of this chapter. As I will describe below, they find consolation in ways that testify to a discrepancy between what official religion prescribes and how they actually live their religion. This state of affairs also raises an important, practical question with regard to the future planning of churchyards: for whom will they be landscapes of consolation? I will address this question towards the end of the chapter, but first we need to understand how the material side of Protestant religion manifests itself in Danish graveyards.

The seemingly secularised look of Danish churchyards

Religious practices in connection with grave visiting in contemporary Europe have been under-researched, with Bailey (2006) as a prominent exception. This is surprising since death is seen as fundamental to religion in several classical studies on religion, and burial grounds therefore seem to be an obvious place to look for religion (see Berger 1967). One of the reasons for this neglect is probably the common expectation that with modernisation, a secularisation of all burial grounds has taken place in Europe. There have, however, been different routes to modernisation, and the modernisation of burial grounds during the last two centuries did not always imply institutional secularisation. Denmark is a case in point. As I shall describe in more detail later, the Lutheran Church played an active role in the modernisation of the layout of graveyards and even embraced cremation. What is more, the Church not only controls the great majority of burial places but also most crematoria. Tony Walter has thus differentiated between a municipal, a commercial and a religious funeral model and places the Nordic countries, including Denmark, in the last category. But although not formally secularised institutionally, Walter (2005) nevertheless considers the Nordic churchyards to be *de facto* secularised.

Broadly speaking the churchyards in Denmark have indeed been subject to a great deal of change in the twentieth century. The change in terms of material culture, admittedly, seems to suggest a process of secularisation. Cremation did become very popular and turned out to be a significant development affecting the structural design of graveyards. One of the most prominent changes resulting from the increase of cremations has been that almost all Danish churchyards have been extended with new lawn sections with urn graves on which only small, flat gravestones of a uniform size are allowed (Sørensen 2010). The

Danish language. In English the word 'cemetery' tends to refer to secular burial grounds owned by the municipality or private business. Such cemeteries hardly exist in Denmark. One regularly finds a more liberal attitude towards an 'aberrant' material culture in the burial grounds run by Danish municipalities.

stones have no religious inscriptions or symbols but only names and dates of birth and death. Furthermore, the fact that coffins have been replaced by urns implies an annihilation of the religious symbolism of the inhumed body facing east, awaiting the second coming of Christ (Jupp 2005).

These aspects of 'secularisation', however, cannot be taken at face value, due to the way Protestantism relates to material culture. In order to fully understand this, we first have to look at how a new Protestant view on material culture emerged at the time of the Reformation, in opposition to the common practices of Catholics, which had become suspect.

The reformation of material culture

Protestantism distanced itself at the outset from the material side of religion and contested Roman Catholic material culture. Violent iconoclasms, and intense theological debates about the sacrament of the Holy Communion, all centred on the view that the only thing that could make God present was the word of God. This was the only contact point between humans and God, and the only way for humans to reach this point was through spiritual belief; no aspect of the material world could be used as intermediary. Yet this stance is hard to sustain, because in actual practice material expressions of religiosity are indispensable. As Arweck and Keenan point out, 'The idea of religion is largely unintelligible outside its incarnation in material expressions' (2006: 2–3). An absent God must somehow be made present and accessible in this world, and material culture plays a vital role in this connection.

Figure 6.1 Askov Kirkegård, 2014. A typical Danish churchyard with traditional graveplots surrounded by hedges and a new lawn section for urns.

In other words, we are dealing with Protestant material religion after all. In addition, the conflicts about material culture, dating back to the Reformation, demonstrate that Protestants, despite their professed split between matter and spirit, in fact attributed considerable importance to material things: how could Roman Catholic material culture be dangerous and misleading if it did not mean anything? To put it slightly differently, we need to make a distinction between Protestant self-representations that downplay the role of materiality on the one hand, and actual religious practices that cannot escape some kind of material expression on the other (Meyer and Houtman 2012: 12).

The downplaying of the role of material religion – or 'Protestantism's dematerializing inclination' as Meyer and Houtman (2012: 15) call it – tends to render the material side of Protestantism almost invisible, even though it is still present, as I have pointed out above. This inclination does not, however, only exist within Protestantism; it has also been prevalent amongst scholars of religion. Tending to take Protestant understandings of religion for granted and privileging spirit over matter, belief over ritual, and inward conviction over 'mere' outward action, we have not been looking for and investigating the material side of Protestant religion because this side was not counted as part of 'real' religion. Instead, we have erroneously taken a *de facto* Protestant material culture for a secular one. The related view, that the appearance of Danish churchyards would be solely a matter of aesthetics, having nothing to do with religion, is equally mistaken.

The displacement of consolation

Within the prevailing continuing bonds theory on grief it is a dictum that 'if death ends a life it does not end a relationship' (Goss and Klass 2005: 5). It has been pointed out that many mourners find resolution to their grief by integrating the dead in their continued life, and Klass (2014: 6) accordingly states that 'Solace alleviates, but does not remove distress'. This means that loss is not something we 'get over' by 'working through' the grief and 'letting go' of the dead as believed within the Freudian grief paradigm (Goss and Klass 2005: 4f.). That the bereaved should sever their bonds with the dead is, however, not a recent idea; it can also be found in earlier historical periods (Goss and Klass 2005: 8), though 'the most common consolation in the histories of religions comes within continuing bonds with the dead' (Klass 2014:15). Within Christian theology contradictory understandings of the relation between the living and the dead thus came to the fore at the time of the Reformation (Goss and Klass 2005: 219–221), and as I shall explain, the split between Catholic and Protestant understandings also affected the churchyard as a landscape of consolation in Denmark, when it became Lutheran in 1536.

Before the Reformation, the grave was a source of consolation. On this side of the grave, the living could interfere in what happened on the other side of the grave, and from the other side, the dead (saints) could interfere in what happened on this side. According to Roman Catholic theology the dead were still present in the sense that the bonds between the dead and the living continued. The living could acquire indulgences, light candles, pray and have masses said for the souls of the dead to shorten their time in purgatory. Some venerated dead, the saints, could also interfere in the world of the living and help them. Material objects, especially relics, played an important role in creating these bonds and in giving the absent dead a presence. Together with the shrines of saints they formed important contact points between the living and the dead.

With the Reformation, the world became much smaller, because the Protestant world only consisted of the living. Reformers not only contested the way in which the Roman Catholic Church made the absent God present in this world by means of mediators and material objects as was the case, for instance, in the conflict about the Holy Communion and the Roman Catholic doctrine of transubstantiation. The presence of the dead was also an important issue in this religious conflict. Indeed, discussions about the teachings concerning purgatory triggered the conflict between Martin Luther and the Roman Catholic Church, leading to his excommunication in 1521. As a consequence of the Protestant doctrine of justification by faith alone, it was God alone who dealt with the souls of the departed and no one else could intercede. To be dead meant to be sleeping in the hands of God until Judgment Day; no living person could reach out to the dead and vice versa (cf. Kroesen 2014).

So while death in the Catholic context was a process in which the fate of the souls of the dead could be influenced by the Church and the bereaved, the living and the dead became fully separated in the Protestant context. In other words, death was a boundary that could not be crossed (Koslofsky 2000; Gordon and Marshall 2000). The relationship between the living and the dead henceforth remained confined to the memory of past lives and the eschatological hope of salvation. Consolation had to be found in the belief in the word of God and His mercy. This meant that there was no longer any positive theological interest in the place of the dead in itself. Instead, the prime theological interest concerning graveyards was to keep them free from any 'Catholic' bonds between the living and the dead.

Regulating consolation in Lutheran Denmark

While the bonds between the living and the dead had been theologically severed at the time of the Reformation, it subsequently had to be ascertained that they were broken in practice as well. No communication between the living and the dead was to take place anymore. Graves were only to be places of memory and eschatological hope – consolation had to be found in the right way.

One way to ascertain this, found within the ranks of more radical reformers inspired by Calvin and Zwingli, was simply to destroy the material culture that could manifest continued bonds between the living and the dead. Thus, gravestones and bone-houses were smashed, graves were levelled and candles placed on the graves were also removed (Sörries 2009: 118). Reciprocity between the living and the dead was out of bounds. Marcel Mauss' essay on the gift illuminates the underlying law of reciprocity. When someone gives a gift, the receiver is obliged to donate something in return – just as a question expects an answer in reply (Mauss 1954). In the same way, placing things on graves can be seen as gift-giving to the dead. And since they are the postulated recipients, agency is attributed to the dead, based on the understanding that they might return the favour. The dead are somehow conceived of as partners in this exchange and dialogue with the bereaved donors. The grave with the deceased's remains happens to be a site for communication with the dead *par excellence*; the placing of things on the grave and/or talking to the dead makes it possible for the living to maintain continued bonds with the dead. These practices accord a presence to the absent dead (All Souls' Day being a prime example of this in the Catholic liturgical calendar). This sort of communication would of course be in conflict with Protestant beliefs, and it was in order to prevent this that some radical reformers found the solution described above, in simply making the graves and/or the things placed on them disappear.

Luther's stance was more complex. He did not consider things in themselves a problem; they were soteriologically indifferent (adiaphora). For him the problem was instead the way in which people related to things; this was what had to be changed. This also applied to churchyards (Sörries 2009 101; Illi 1992: 111). Following Luther's theology, warnings about how to visit the grave properly were issued early on in Denmark. Peder Palladius, the first bishop of Zealand after the Danish Reformation in 1536, describes how he inspected his diocese in the 1540s and warned people that they were not to visit the graves in order to pray for the dead, but only to remember that they themselves would die one day and that they would be judged too (Lausten 2003: 31–32). The Lutheran way of ascertaining that communication between the living and the dead would not take place was thus less opposed to material culture than the stance of reformed Protestants. If material things were used to manifest memories or eschatological hope, they were acceptable.

More radical views on the material culture of the churchyard, however, became present in Denmark at a later stage. Having visited Moravians in the Dutch city of Zeist, the Danish king Christian VII allowed the Moravians in 1772 to build their own town in Denmark. It was to be named Christiansfeld and right from the start a churchyard, 'the God's Acre', was established (Bøytler and Jessen 2005: 24–27, 177–189). It became a true copy of the one of the mother colony in Herrnhut, in Saxony, that in turn looked very much the same as typical churchyards of reformed Protestants in Germany (Sörries 2009 119, 124–126). All graves on the God's acre in Christiansfeld are placed in regular rows at a fixed distance to one another and are marked by identical flat stones, engraved with only a number, name, place and dates of birth and

Figure 6.2 The Moravian graveyard 'God's Acre' in Christiansfeld, a UNESCO heritage site since 2015. The equality of humans in the eyes of God is manifested by the similarity of the graves.

death, aiming to symbolise the equality of all humans in the eyes of God. And, very importantly: there was to be no other decoration. Today, restrictions have been eased a bit, and it is now allowed to plant flowers, but only one-season flowers that will disappear quickly (Bøytler 2001: 29–34). Clearly, clear-cut restrictions were placed on the material culture of this churchyard, to prevent it from materialising continuing bonds between the living and the dead – again, consolation was instead to be found in the word of God.

As I will show, it was the churchyard in Christiansfeld, with its specific Protestant norms of consolation, which was to become one of the main inspirations for the design of churchyards in Denmark in the twentieth century. But because the development of churchyards in this period was highly influenced by the advent of cremation, we will first have closer a look at this development and see how Protestant material religion played a role here and how the Lutheran Church joined forces with what was regarded as progress and modernisation.

Cremation and Protestant heritage

Modern cremation only became technically possible as late as the second half of the 1870s, but already in 1881 the first cremation society was founded in Denmark. By 1886, it had built the first Danish crematorium in Copenhagen and the first cremation was conducted. This placed Denmark among the very first countries to make use of this new technology of bodily disposal. Until then modern cremation had only taken place in crematoria in Italy (since

1876), Germany (since 1878) and the United Kingdom (since 1885) (Davies with Mates 2005). The Danish authorities, however, reacted strongly against it: a police ban was issued against the use of the crematorium. It was upheld by two court decisions until 1892, when cremation was finally legalised by the government (Larsen 1996). There were also strong reactions against cremation from the public, especially some currents within the church fiercely opposed to it. But the church was divided and never formed a united opposition (Møller 2007). It turned out to be difficult to find substantial theological arguments against cremation from a Lutheran point of view, and over time the part of the church that had no problem with cremation got the upper hand.

In the same year when the first cremation took place in Denmark, cremation was banned by the Roman Catholic Church. The ban was rescinded in 1963, as part of larger reforms at the time of the Second Vatican Council (1962–1965). What made cremation more easily acceptable to many Protestants than to Roman Catholics was the previously described Reformation split between matter and spirit. This split had become even more accentuated within Protestantism with the rise of natural science and the materialist critique of religion; it made Protestant theology retreat even further from the material world to the inner world of the spirit (Meyer and Houtman 2012: 6). The challenges to the belief in a bodily resurrection offered by the natural sciences made it increasingly attractive and common for Protestants to 'spiritualise' their afterlife beliefs in the nineteenth century. Belief in the immortality of the soul came to have a much more prominent role, at the expense of the belief in a bodily resurrection (Badham 2005). With this, the dead body lost its religious importance more and more and the symbolic placing of the body facing east, awaiting the second coming of Christ became empty. The dead body became more like a disposable thing and as such it could be cremated. Whether burial or cremation was chosen was considered religiously a matter of indifference, since for Lutherans salvation did not depend on outer forms (Rald 1936).

The transformation of the dead into ashes made it possible to place the remains of the dead in new ways without causing hygienic problems. The ashes could be placed above the earth as well as be interred, did not have to be placed in a churchyard but could just as well be placed somewhere else and finally cremation also made it possible to divide or spread the ashes. In the earliest history of cremation in Denmark, however, ashes were disposed of in only one way: ashes were put into urns, and the urns into columbaria (Foreningen Liv & Død n.d.). This was also mainly the case in e.g. England and Germany (Grainger 2005; Fischer 2009) and an interesting feature of the early development of modern cremation. Cremationists saw themselves as modernisers who put reason above tradition and in this vein, typical arguments in favour of cremation were that it was hygienic and efficient, in the sense that cremation freed space that the living could make better use of. However, at the same time cremationists clearly reinvented the old Roman tradition of the columbarium to legitimise the new disposal technique. The same reinvention of old Roman tradition is also visible in the use of a classical style in many of the early crematoria (Jensen 2002: 173–179; Kragh 2003: 224).

However, the crematoria built in Denmark from 1926 onwards had a new appearance and came to look like small chapels. And in fact they were. While the early crematoria had initially been built by the Cremation Society and later on by local municipalities, the Church now began to get involved. The new involvement of the church resulted in a genuine building boom. While only three crematoria had been built between 1886 and 1925, sixteen so-called chapel-crematoria were built in the next fifteen years (Markussen 2011). This was a very visible sign that the Lutheran Church in many places by now not only accepted cremation but embraced it. Full equality of burials and cremations, however, was only reached as late as 1975. Previously, individual priests could refuse to take part in a funeral service that would be followed by cremation if it would be against their conscience (Kragh 2003: 224).

In the process that led to the rapid increase in the building of crematoria, the cremationists also changed signals. An anti-clerical wing did exist within the cremation society that united members of both bourgeois and social democratic background. This wing had strong connections with the Labour movement and had its origins in the fight that led to the legalisation of civil funerals in 1907 (Kragh 2003: 222–223). But there were other goals for them to consider than just fighting the church. A certain number of cremations was needed to keep the crematoria in operation, since otherwise cremations would remain very costly, and this clearly went against the Social Democratic ambition of making cremation available to all citizens. The cremation society, therefore, had to recruit more members, and with the near-monopoly of the Lutheran church, this was only possible by recruiting church members (Rald 1936).

In consequence, the cremation society ended up being careful not to take an atheist-materialist stance that could offend Christians and underlined instead that cremation was religiously neutral (Secher 1956: 76). It was merely an outer form and it did not in any way exclude or affect a religious understanding of death. With overlapping views on materiality as adiaphorous, church and cremation society could join forces. In Denmark, it was henceforth possible to feel both modern-cum-rational and Christian, while these stances became opposites in many other countries. Today the Lutheran Church thus owns two-thirds of Danish crematoria, whereas crematoria are seen as features of secularisation elsewhere. The Danish development therewith exemplifies that modernisation did not always exclude the church.

The incorporation of the new cremation technology into the Christian tradition was visible in the disposal of ashes, even earlier than in the style of the crematoria. Whereas ashes were placed in columbaria in the early history of cremation, in 1910 it became legal to inter ashes in churchyards (Foreningen Liv & Død n.d.), and over time various types of graves for urns were developed that became much more popular than the columbarium (Andersen 1972). During the twentieth century, these new forms changed the material culture and appearance of Danish churchyards dramatically, because cremation in Denmark not only became legal and possible but, as mentioned above, also very popular. Denmark was the first Western European country after the United Kingdom to reach a cremation rate of 50 per cent in 1976 (Davies and Mates 2005), and according to the latest cremation statistics from 2017, Denmark today has a staggering cremation rate of 83 per cent (Danske Krematoriers Landsforening 2017). Rather than seeing this growth in the number of cremations as being caused by secularisation and religious norms becoming less influential, it must be understood as the opposite: as a result of the Protestant split between matter and spirit resting on a denial of the significance of the material world to get into contact with God. Seen from that angle it comes as no surprise that modernisation and cremation were embraced in Denmark, a society so much imbued with Protestantism.

The new-old deathscape and the strengthened church

One of the most important changes that followed in the wake of the rapid growth of cremation is that most Danish churchyards have now been extended by adding lawn sections with urn-graves, as described in the beginning of this chapter. And it is especially on these lawn sections that the influence from the Morovian churchyard in Christiansfeld is most visible. The identical flat grave-stones placed in rows, at the same distance from each other and engraved with only names and dates, mirrors the Morovian churchyards strikingly. And that is no coincidence. When such a lawn section was designed for the first time, in 1945–1950, by G. N Brandt on the famous Mariebjerg Kirkegård in Gentofte, it was in fact done with explicit inspiration from Christiansfeld (Sommer 2003: 240–241). What looks like a secularised churchyard to Tony Walter thus turns out to express a heavy Protestant heritage. So does another new feature that developed in twentieth-century churchyards in Denmark: regulations of what – or rather what not – to place on the graves. This feature we have also already met on the Morovian churchyard in Christiansfeld. And again, it was the architect G. N. Brandt who introduced it. Already in 1922, he had argued for an extended use of statutes and put his views into practice when he began developing Mariebjerg Kirkegård from 1925 onwards. He considered the use of statutes necessary to be able to strike an aesthetic balance between individual and collective interests (Falmer-Nielsen 2002).

G. N. Brandt together with his colleague J. Exner became key figures in churchyard-landscaping in Denmark in the twentieth century. J. Exner also saw the churchyard in Christiansfeld as the ideal and echoed Brandt's view on the need to introduce statutes for aesthetic reasons (Exner 1961: 35–43, 56b). The God's Acre's theological accent on equality was of course also politically attractive in the context of a Social Democratic welfare state. As a result, Danish churchyards today, besides having a widespread use of lawn sections, are extremely regulated. Different areas of the churchyards have different rules about what one is allowed to place there. In newer sections people are often formally forbidden to place things on the graves, with the exception of fresh-cut flowers. Sometimes the statutes can even be very specific in mentioning things that are forbidden to be put on the graves: figurines of animals, gnomes and santas, benches, electric light and sometimes even candles. On the traditional grave plots people are normally still relatively free to place what they want within the boundaries of not disturbing the order and peace of the churchyard, but what is deemed befitting in this respect is up to the local parish council to decide (Kirkeministeriet 1996).

The fact that the rules are stricter on the newer parts of the churchyards than on the older parts shows that over time the Lutheran Church has assumed much more control over the appearance of the graves. This has happened at the expense of the family. That the family has lost control over the dead forms part of a general process of the professionalisation in dealing with death, where doctors came to take care of the dying, undertakers of the funeral and new professionals employed on the churchyards for digging and maintaining the graves instead of family, friends and neighbours (Walter 1996). Because the secularisation of churchyards in modern times is taken for granted, it has been generally overlooked that this modernisation process could also include the church, namely in countries with a religious funeral model. Due to the Lutheran Church's increased control of the graves it in fact seems more apt to talk about a sacralisation than a secularisation of Danish churchyards in the twentieth century.

The Lutheran Church got a particularly strong grip on disposal due to its near-monopoly on burial grounds. This near-monopoly could easily have been broken with the advance of cremation. As mentioned, ashes could be placed anywhere without causing hygienic problems. But as in Germany, ashes still have to be disposed of in churchyards. It can be imagined that the Freudian grief paradigm prevailing in the twentieth century, focusing on the need for detachment between the living and the dead, might have been important to keep up what Germans call *Friedhofszwang*. The only exception from this duty is that it is possible to spread the ashes, but this can only be done on open sea and it has to be made plausible that it is in accordance with the wish of the deceased.

Contesting Protestant norms of consolation

Just as radical Protestant reformers made unwanted things disappear from graves in the sixteenth century, so have the formal statutes introduced to Danish churchyards during the twentieth century, and efficiently implemented by the new churchyard-professionals, been capable of making religiously unwanted things disappear. What is considered to be the 'right' aesthetical balance between collective and individual interests is far from religiously neutral. The present material culture of Danish churchyards is very uniform and there are not many material items to be found on graves compared with other countries. Admittedly, this has been achieved in a less violent way than at the time of the Reformation, but at stake, I argue, is still the same distinctive Protestant notion of consolation. Along with this comes a strong suppression of material culture of non-prescribed forms of religiosity and consolation, although covered up in aesthetic arguments resulting from Protestant downplaying of the material side of religion. It is this overlooked repression to which I will turn in the remaining part of this chapter.

Since the beginning of the twenty-first century there has been a rise in conflicts about what items of material culture are to be considered proper in churchyards. Fully in line with G. N. Brandt and J. Exner these conflicts are mainly understood in terms of individual taste versus interests of the collective. Underneath the varnish these conflicts are, however, about much more than different views on aesthetics. They are fundamentally triggered by different theologies of death and express other views on consolation than the official Lutheran standpoint does. It is the silencing of these other views that is the real kernel of the conflicts, as the following case, which I have followed through case files of Aarhus Diocese and newspaper articles, makes clear.

The archetypical conflict dates back to 2001 and ushered in a series of similar conflicts. The conflict began when, around Christmas time, a mother placed a cross, decorated with fir branches and an electric light chain, on the grave of her recently deceased teenage daughter. This act, however, soon turned into a conflict with the local parish council in Grenaa, which did not find electric light to be appropriate in the churchyard. The council ordered the mother to remove the electric light chain and stated that if she were not willing to do so, the council would have it removed. The mother complained about the decision of the parish council, and the conflict was addressed on both the level of diocese and government and received a lot of public attention before it was settled in 2005. The mother was given a dispensation by the parish council to have a light chain on the grave, but only between the 23th and the 30th of December, solely on one cross and with a maximum height of 40 cm (instead of 150 cm).

When the parish council forbade the mother to place an electric light chain on the grave of her daughter, it explained its decision on aesthetic grounds. In a letter from the parish council to the diocese, the council explicitly declared that they considered this decoration to be inappropriate and further explained that they were afraid it would turn the churchyard into a 'Tivoli' (referring to the famous amusement park and pleasure garden in Copenhagen) and would distress other visitors on the churchyard because it was not part of the tradition of material culture on Danish churchyards. But that there was something more at stake than aesthetics became clear from the parish council's letter seeking to explain why electric lights on graves were problematic, while candles were not: the council wrote that candles suited the remembrance of a dead person and that the same was not the case for electric light. It thus seems that the real problem behind the rhetoric of aesthetics was that the mother did not relate to the electric light chain in the right way. When visiting the grave, she apparently was doing something other than just remembering her dead daughter.

That the heart of the matter was about how one should relate to grave decorations and about the right religious way of finding consolation, becomes clear when looking at the explanations the mother gave to a newspaper about why it was so important for her to have this electric light on the grave: 'I can buy toys and clothes for my other children. But the only thing I can give my daughter is a light in the darkness' (*BT*, 28 September 2002). To this mother her dead daughter obviously was not a memory of the past as prescribed by Lutheran theology. On the contrary, the deceased daughter was someone with whom the mother could still interact and to whom she could give things. The mother's continuing bonds with her daughter were made manifest in a material way by placing a Christmas decoration with electric lights on the grave.

Official religion versus lived religion

Despite the rising number of public conflicts and contestations of the Protestant norms of consolation permeating Danish churchyards, they are still relatively rare. However, publicised conflicts are only the tip of the iceberg. Most conflicts are managed on local churchyards without attracting attention. When statutes explicitly mention specific objects like the figurines of animals, gnomes and santas, benches and electric lights, it testifies to the fact that the placement of such objects has caused conflicts. But the practice of maintaining continuing bonds with the dead, I would argue, might be even more widespread than indicated by these regulations. A silent subaltern existence is another solution to repression, instead of open conflict. In this respect, it is remarkable that the head of the churchyard committee of the parish council in Grenaa made the following comment to a newspaper: 'Why not just do it at home [putting up an electric light chain]. It does not have to be demonstrated that publicly' (*BT*, 28 September 2002). The comment suggests that this key member of the Grenaa parish council knows full well that interactions between the living and the dead are a fact of Danish culture, but prefers them to take place outside the churchyards, in a non-public setting.

That this is in fact the case is demonstrated by the Danish Internet memorial mindet.dk. Text messages left on its pages testify to an ongoing communication with the dead, since people not only write about their memories of the dead person in question but also directly address the dead person. This is also known from other research (see e.g. Roberts 2004; Kasket 2012). However, it is overlooked that Internet memorials also describe practices on churchyards. Users upload pictures that document the grave in its various stages, freshly made and covered with flowers, the headstone and ornaments placed, and so forth. They also document grave visits and often write about their visits and their practices at the grave. There are many photos of grave visits on special days, such as the dead person's birthday, the wedding anniversary, Christmas and Easter. These are the days when the whole family normally would meet if the dead person would still have been alive, and they still do meet, but now in the churchyard. A good example shows the scene of a typical Danish birthday celebration, complete with the Danish flag, coffee and the favourite cake of the birthday-'child', only that this celebration takes place at the graveside and is photographed by the widow. The caption reads: 'Some of your children, children-in-law and your grandchild congratulate you [with your birthday]'. Clearly, the dead person is addressed (mindet.dk, posting on 8 November 2008). The mother from the Grenaa-case is certainly not alone in maintaining continuing bonds with a dead loved one manifested through things placed on the grave. Mindet.dk in this way gives clear evidence that non-public interaction between the living and the dead is not something that solely takes place outside churchyards, as the remark from the head of the Grenaa parish council seemed to suggest. It also regularly takes place in the churchyards.

That the grave site is an important source of consolation, at odds with Protestant theology, is thus a widespread phenomenon, also for Lutherans. For instance, the mother in the Grenaa-case was a member of the Lutheran Church. And since mindet.dk is owned by the only Christian newspaper in Denmark, it must also be assumed that most of the people contributing to the website are affiliated with the Lutheran Church. To explain the discrepancy between the theologically prescribed relations between the living and the dead on the one hand, and what is practised on the other hand, an apparently attractive conclusion would be that the Lutheran Church has lost authority. People maintaining continuing bonds with the dead cannot be counted as 'genuine' Lutherans, even though formally being members of the Lutheran Church. This conclusion would be in line with findings by researchers such as sociologist Phil Zuckerman (2008). In his book *Society without God* he describes Denmark as one of the most secularised countries in the world, while attributing a membership rate of the Church as high as 75.9 per cent (Kirkestatistik 2017) to mere tradition.

That this seemingly attractive conclusion is inadequate, is exemplified by the Grenaa-case. The mother in question considered herself to be a good Christian in spite of having a relationship with her dead daughter, at odds with Lutheran theology. And as she later was elected as a member of the parish council, it is clear that others also considered her a good Lutheran (*Kristeligt Dagblad*, 11 November 2004). We have to be aware that religion is always lived and different from what is prescribed, even when it comes to committed Christians (Stringer 2008: viii). Textbook religion does not exist in real life. Religiously highly committed Lutherans might well find consolation in what must be described as decidedly non-standard ways, when interpreted against the background of a textbook of Lutheran dogma. With this in mind the rising level of conflict between what is theologically prescribed and how things actually take place is better explained as a result of how the Church has expanded its grip on the churchyard and left less space for family members to find consolation in divergent ways.

Conflicts about what one is allowed to place on graves are not just about aesthetics, they are also about people fighting to practise their religion within the framework of the Church, in ways that make sense with regard to their own life and that offer consolation. What is meaningful to people might not be so much dogmas and beliefs, as one is led to believe by Protestantism, but rituals and practices (McGuire 2008: 19–44). In this connection, it is interesting that photos depicting baptism are repeatedly found amongst the photos uploaded on mindet.dk. It seems that to these people the shared ritual of baptism is what establishes the possibility of having continuing bonds with the dead. So while the Protestant theological rejection of continuing bonds between the living and the dead is deemed not to be religiously meaningful, the ritual of baptism clearly is.

Planning for future consolation

It is to be expected that the regulation of material graveyard culture and its implicit Protestant norms can make it difficult to find consolation for non-members of the Lutheran Church. This group has been growing, mainly due to immigrants and refugees coming to Denmark. Before the 1960s, Denmark was a more or less monocultural and mono-confessional country. Therefore, discussions, legislation and planning of churchyards have mainly focused on how to include atheists and people with other religious backgrounds. The Church Ministry, for instance, published a new guideline, entitled *Kirkegården – begravelsesplads for alle* (The churchyard – burial ground for everyone), in 1996, urging parish councils to show a special openness and responsiveness to these groups, given the near-monopoly of the Lutheran Church on burial grounds.

That the churchyards' immanent Protestant norms of consolation can be problematic is thus recognised. However, it has not been problematised that Lutherans can also have problems with finding consolation in the prescribed way, as I have shown here. If the churchyards are to be for everyone, should they then also be places where everyone can find consolation, including Lutherans who find consolation in non-dogmatic ways? Would it be appropriate if discussions, legislation and planning of future churchyards took into account the needs of these church members too?

Many people – also Lutherans – find consolation in maintaining continuing bonds with their dead by placing things on the grave. Not only for aesthetic but also for normative religious reasons, graveyard regulations appear to present obstacles. The statutes are often violated and based on this experience parish councils have been urged to explicitly inform about statutes when families are choosing graves to avoid future conflicts about grave decorations (Falmer-Nielsen 2002). If these conflicts were just about aesthetics, more information might be a good answer. But as I have shown here the conflicts are also about different theologies of death and consolation. A planning based on what is practised would give less reason for conflict and would result in churchyards being more satisfying places to find consolation for Lutherans and non-Lutherans alike, no matter how they live their religion.[3]

3 An earlier version of this chapter appeared in *Nederlands Theologisch Tijdschrift* 68/1&2 (2014), 101–119.

References

Andersen, A. 1972. Forandring i jordfæstelsesskikke statistisk belyst. *Vore kirkegårde* 24/3, 17–28.

Arweck, E. and Keenan, W. 2006. Introduction: material varieties of religious expression, in *Materializing Religion: Expression, Performance and Ritual*, edited by E. Arweck and W. Keenan. Aldershot: Ashgate, 1–20.

Badham, P. 2005. Soul, in *Encyclopedia of Cremation*, edited by D. J. Davies with L. H. Mates. Aldershot: Ashgate, 376–377.

Bailey, E. I. 2006. *Implicit Religion in Contemporary Society*. Leuven: Peeters.

Berger, P. 1967. *The Sacred Canopy: Elements of a Sociological Theory of Religion*. Garden City, NY: Doubleday.

Bøytler, J. 2001. Gudsageren Christiansfeld. *Kirkegårdskultur*: 29–34.

Bøytler, J. and Jessen, J. T. 2005. *Christiansfeld: Livet og husene*. Søborg: udgivet af Det Danske Idéselskab.

Danske Krematoriers Landsforening. 2017. Statistik. http://www.dkl.dk/info/statistik

Davies, D. J. with Mates, L. H. 2005. Cremation statistics, in *Encyclopedia of Cremation*, edited by D. J. Davies with L. H. Mates. Aldershot: Ashgate, 431–456.

Exner, J. 1961. *Den danske kirkegård og dens problemer*. Horsens: GAD.

Falmer-Nielsen, M. 2002. Servitutter – pro et contra. *Kirkegårdskultur*: 34–40.

Fischer, N. 2009. Aschengrabmähler und Ashenanlagen der modernen Feuerbestattung im späten 19. und frühen 20. Jahrhundert, in *Grabkultur in Deutschland*, edited by R. Sörries. Berlin: Reimer, 151–161.

Foreningen Liv & Dødn.d. (earlier known as Dansk Ligbrændingsforening/The Danish Cremation Society). *Ligbrændingens historie, 1910–20*. http://livogdoed.dk/temaer/ligbraendingens-historie/

Gordon, B. and Marshall, P. 2000. *The Place of the Dead: Death and Remembrance in Late Medieval and Early Modern Europe*. Cambridge: Cambridge University Press.

Goss, R. E. and Klass, D. 2005. *Dead but Not Lost: Grief Narratives in Religious Traditions*. Walnut Creek: AltaMira Press.

Grainger, H. 2005. Columbaria, in: *Encyclopedia of Cremation*, edited by D. J. Davies with L. H. Mates. Aldershot: Ashgate, 128–129.

Illi, M. 1992. *Wohin die Toten gingen: Begräbnis und Kirchhof in der vorindustriellen Stadt*. Zürich: Chronos.

Jensen, J. F. 2002. *Vest for paradis: Begravelsespladsernes natur*. København: Gyldendal.

Jupp, P. 2005. Resurrection and Christian thought, in: *Encyclopedia of Cremation*, edited by D. J. Davies with L. H. Mates. Aldershot: Ashgate, 353–358.

Kasket, E. 2012. Continuing bonds in the age of social networking: Facebook as a modern-day medium. *Bereavement Care* 31/2, 62–69.

Kirkeministeriet (1996). *Kirkegårdsvedtægter: en vejledning for menighedsråd, kirkegårdsbestyrelser og provstiudvalg*.

Kirkestatistik (2017). Folkekirkens medlemstal. *Kirkeministeriet*. http://www.km.dk/folkekirken/kirkestatistik/folkekirkens-medlemstal/

Klass, D. 2014. Grief, consolation, and religions: a conceptual framework. *Omega: Journal for Death and Dying* 69/1, 1–18.

Koslofsky, C. M. 2000. *The Reformation of the Dead: Death and Ritual in Early Modern Germany*. Basingstoke: Palgave.

Kragh, B. 2003. *Til jord skal du blive. Dødens og begravelsens kulturhistorie i Danmark 1780–1990* (Skrifter fra Museumsrådet for Sønderjyllands Amt 9). Sønderborg: Aabenraa Museum.

Kroesen, J. 2014. De slaapdoodmetafoor in de Groninger grafkunst. *Nederlands Theologisch Tijdschrift* 68/1–2: 52–66.

Larsen, C. 1996. Københavns første ligbrænding. *Siden Saxo* 13/3: 31–36.

Lausten, M. S., ed. 2003. *Peder Palladius: En Visitatsbog.* København: Anis.

Markussen, A. 2011. Da Sønderjylland fik et krematorium: kremeringens fremvækst i sampil med kirke, kristendom og velværdsstat. *Sønderjysk Månedsskrift* 1, 3–12.

Mauss, M. 1925/1954. *The Gift: Forms and Functions of Exchange in Archaic Societies.* London: Cohen & West.

McGuire, M. 2008. *Lived Religion: Faith and Practice in Everyday Life.* Oxford: Oxford University Press.

Meyer, B., and Houtman, D. 2012. Introduction: material religion – how things matter, in *Things: Religion and the Question of Materiality*, edited by B. Meyer and D. Houtman. New York: Fordham University Press, 11–23.

Møller, A. S. 2007. Brændes eller begraves? *Fortid og nutid* 2, 83–102.

Rald, N. J. 1936. Kirken og ligbrændingen. *Dansk Ligbrændingsforenings Aarsskrift.*

Roberts, P. 2004. The living and the dead: community in the virtual cemetery. *Omega: Journal of Death and Dying* 491, 57–76.

Secher, K. 1956. *Ligbrænding i Danmark.* København: udgivet af Dansk Ligbrændingsforening.

Sommer, A. 2003. *De dødes haver. Den moderne storbykirkegård.* Odense: Syddansk Universitetsforlag.

Sørensen, T. F. 2010. A saturated void: anticipating and preparing presence in contemporary Danish cemetery culture. *An Anthropology of Absence: Materializations of Transcendence and Loss*, edited by M. Bille, F. Hastrup and T.F. Sørensen. New York: Springer, 115–130.

Sörries, R. 2009. *Ruhe sanft. Kulturgeschichte des Friedhofs.* Kevelaer: Butzon & Bercker.

Stringer, M. D. 2008. *Contemporary Western Ethnography and the Definition of Religion.* London: Continuum.

Walter, T. 1996. *The Eclipse of Eternity: A Sociology of the Afterlife.* London: Macmillan.

Walter, T. 2005. Three ways to arrange a funeral: mortuary variation in the modern west. *Mortality* 10/3: 173–192.

Zuckerman, P. 2008. *Society without God: What the Least Religious Nations Can Tell Us about Contentment.* New York: New York University Press.

Part III
Beyond the Global North

7 Moving through the land

Consolation and space in Tiwi Aboriginal death rituals

Eric Venbrux

Introduction

The Tiwi Aborigines of North Australia have a cycle of mortuary rites in which territorial passage, wailing and consolation are interlinked. Ideally, the rites start in the localities of the living and with intervals go on in space and time until the beginning of the final rite at the burial place, an area reserved for the spirits of the dead. In the ritual drama, given its purpose to direct the spirit of the dead from the world of the living to the world of the dead, the people of different bereavement status all play a role in the remembrance and dissolution of a particular metaphorical relationship with the deceased. In this context compassionate support and protection are given to the bereaved. This chapter focuses on the emotional geography of the Tiwi death rites, the dialogues with the deceased in wailing and laments, and the ways in which the ritual drama gives consolation.

Consolation has been given far too little attention in the anthropology of death, let alone the link between consolation and place or space. According to Klass (2013: 610), 'To be consoled is to be comforted. Solace is found within the sense of being connected to a reality that transcends the self.' 'Consolation involves a shift of perspective and an experience of meaning in spite of suffering', in the words of Norberg, Bergsten and Lundman (2001: 544). They make clear that the person in need of consolation due to suffering and the person mediating consolation 'share the suffering in reciprocal presence and availability' (ibid.: 548). As is the case with grief and mourning (Maddrell and Sidaway 2010: 1–2), it 'presupposes space and time' for consolation to take place (Norberg et al. 2001: 550). What is more, as we will see, the Aboriginal *Weltanschauung* is place-centred (Swain 1993; Myers 2002; Venbrux 2015). Stanner has aptly designated this worldview, known as The Dreaming, with the word 'everywhen' (1979: 24). Places can become significant as a result of their identification with the dead (Myers 1986: 134–135; Rose 1992: 69–73; Rose 1996: 70–71; Langton 2002: 260ff). Furthermore, the death rites entail a territorial passage; the transition from the world of the living to the world of the dead being perceived as a journey (Reid 1979: 326; cf. Van Gennep 1960: 153–154). The rites allow the bereaved to express their grief and to receive 'maximum support and protection' (Morphy 1984: 63). This solidarity has been noted in many places in Aboriginal Australia, not only among the Yolngu and the Tiwi (Brandl 1971) in the tropical north, but also among the Martu of the Western Desert: 'Every funeral is well attended and the bereaved have dozens of people giving them comfort and seeing to their needs. Bonds of sentiment and networks of reciprocity ensure that no one grieves alone or without support' (Tonkinson 2008: 49). Durkheim (1995 [1912]) draws heavily on Aboriginal mourning rites to make his argument about social cohesion (Metcalf and Huntington 1991: 33). In these rites, he writes, 'there are tears and laments—in short, the most varied displays of anguished sorrow and a kind of mutual pity that takes up the entire scene' (Durkheim 1995: 400; but see Venbrux 1995: 194–195, 2017a).

This chapter draws on my anthropological fieldwork among the Tiwi from Melville and Bathurst Islands, north off the Australian coast. I revisited my notes from a fourteen-month sojourn in 1988–1989 and five subsequent return visits (in 1991, 1994, 1998, 2002 and 2006) of short duration. One concrete case came to the fore as particularly suitable to illustrate and explain the interrelation between solace and space in Tiwi mortuary behaviour. The case which stands central here shows what happens when the most senior person of a territorial clan dies. If it were possible to follow what happens around the death of an important Aboriginal person, according to Thomson, this would offer the key to a full understanding of 'the whole culture' (cited in Morphy 1984: 127). Because of my good rapport with the bereaved spouse, I believe the present case can also convey something about individual grief (see Reid 1979: 326). Against the backdrop of the ethnographic record on the complex and intricate Tiwi death rites, I seek to foreground the nexus between space and consolation.

In a small-scale, kin-based society such as that of the Tiwi the death of a member has a considerable impact. Indigenous Australians appear to have a life expectancy of ten to twenty years less than other Australians. It is hard to get reliable figures, but the confrontation with death is a very frequent occurrence in the more traditionally oriented Aboriginal societies of North and Central Australia (Glaskin et al. 2008). Tiwi 'mortality rates have been the worst in Australia' (Hoy et al. 2017: 7). From 1986 onwards, their islands saw a wave of suicides, mainly by young men, and were said to have 'the highest suicide rate in the world' (Scott-Clark and Levy 2006). Hoy, Mott and McLeod (2017: 3) count a total of fifty suicides between 1985 and 2010. The number of Tiwi increased from about 1,900 to 2,500 people over that period, which was mainly due to a high birth rate. Robinson (1990) relates the contemporary Tiwi suicides to loss and mourning. Self-wounding is part of Tiwi grieving, but is commonly restrained by consociates. I found it striking that suicidal youths were often said to have visions of the spirit of their dead (grand)father, painted up with white clay and holding a spear; a conventional premonition of death. Cox suggests that suicidal Aboriginal youths consider 'the repeated loss of relatives and friends' as a demise of people who care about them: furthermore, she notes, 'Amongst many of the young people, their socialisation into a world of spirits appeared to produce a feeling of immortality, where death is a loss of a certain physical form and not the loss of life, of self or of agency' (Cox 2010: 251). Is it because they cannot find consolation in this world?

Offering the mourners consolation might be the mortuary ritual's *raison d'être*. The floccinaucinihilipilification of solace in the literature on death rites in Aboriginal Australia – or its sheer neglect – brings into focus the questionable assumption that the sorrow of a considerable number of the performers would be insincere. 'If relatives cry, lament, and beat themselves black and blue, the reason is not that they feel personally affected by the death of a kinsman', according to Durkheim (1995: 400). Osborne (1974: 111n2) even states that when an important Tiwi man dies 'no one feels any real grief, although grief is simulated'. The expressions of grief arise from social obligation or political aspirations. It seems to be rather presumptuous to deny other human beings true feelings. How would one know?

From the sudden shift in the expression of emotion Durkheim seems to infer that the emotions of grief are not authentic: 'If, at the very moment the mourners seem most overcome by the pain, someone turns to them to talk about some secular interest, their faces and tone often change instantly, taking on a cheerful air, and they speak with all the gaiety of the world' (Durkheim 1995: 400). When one realises, however, that Aboriginal people's switching of their expression of emotion involves a shift in focus from one significant person to another one, this behaviour should not be regarded as inconsistent and, therefore, inauthentic. To the contrary, not changing countenance while facing a living person (e.g., a joking friend) would imply a tie-breaking as when directed towards the deceased. The emotions are relational and with the exemption of avoidance relationships Tiwi point their lips towards the person addressed when making verbal statements (Venbrux 1995: 194–195). The European 'concept of the person as a bounded individual' (Bloch 1988: 15) does not apply in the case of these Aborigines. The body is the site in which significant social relations are incorporated, and death of the relative implies the wounding or (imaginary) loss of the respective body part. Simultaneously, body and land are related (Venbrux 1993; Grau 2005). For example, those who have one grandfather in common lose half of their face or have it injured when one of them dies. They find the lost half back or have the face healed at the location of the grandfather's grave, that is, where his spirit dwells. The place not only offers consolation, but is also perceived as empowered by the spirit, promoting health and well-being, including an abundance of food. Furthermore, the burial place is of crucial importance for establishing territorial rights in the surrounding area (Hart 1930a: 173). In the case I am about to discuss the woman who died was seen as the 'big boss' of the northwest of Melville Island.

Connections to country

Tiwi all belong to a specific exogamous matriclan (*imunga*, 'skingroup'), but in the death rituals, as in the rights to land, the focus is on the patrilineal affiliations of the deceased. When the spirit of the dying is supposed to leave the body, it is told in mourning songs to move away towards the burial places of paternal ancestors, or 'dreaming places' (that is the locations of the father's *imunga*: islets off the coast, a waterhole or a rock formation). The funeral and postfuneral rites are focused on directing the spirit of the deceased to those places where the ancestors 'live'. These places, although far away, are addressed and faced when the rites commence. People stamp the ground to call the attention of the spirits of the dead. A goose-feather ball on a cord may be thrown in the direction of the destination, a small rite that also projects the songs and dances across space.

Ritual calls and gestures, together with songs and dances, direct the spirit on a journey along places of totemic significance to its final destination. In the course of the ritual track the deceased's own totems are increasingly given centre stage. If the web of relationships is considered, totemic as well as social ones, the spirit is gradually closed in. This is also expressed spatially. The apotheosis entails a meeting of the living and the spirits of the dead. The latter are kept at bay by the workers, using fighting sticks. In conjunction with this, the final mortuary ritual, the *iloti*, is performed by the spirits and the living actors. At the conclusion of the ritual carved and painted posts are erected around the grave. The posts represent metaphorical bodies, fixing the spirits to the place.

A collective wailing occurs at certain intervals when the distance from the deceased is increased in space and time: when the corpse is put in the coffin, when the dead person is interred, when the deceased is left behind in the graveyard, etc. Participants in the rituals also wail on arrival and they wail at the conclusion of their performances. People strongly identify with their 'country', i.e. one of the eight districts the islands are currently divided into or a part thereof. In wailing, songs, dances and ritual gestures, specific characteristics of, for example, a particular animal or bird abundant in the place in question, are often stressed.

Joan Puruntatameri was the most senior person of the country Munupi. In this she followed her father Tipolei and father's father Tipaklippa. Her father had ten wives (Hart 1928–29: 48). Her mother Tomukareiomau was a daughter of Mangatopi, the main leader of the country where the other township in Melville Island, Milikapiti, was later established. In 1943 Joan married Justin Puruntatameri (Pilling 1958: 75). He was from another part in the same country, where his father's father Korupu had been the leader. Korupu had been an influential man with eleven wives (Hart 1928–29: 2). Justin's mother Tupangkwopeio was a leader in her own right. In the late 1980s Justin was the leader of Mupuni, one of the eight countries represented in the Tiwi Land Council. He was also the undisputed ceremonial leader of Melville Island.

A number of foraging trips with Justin and his family gave me a better understanding of their relationship to their country. Indeed, as Watson puts it with regard to the Walbiri: 'The sensorium and the spatio-conceptual aspects flowing from Aboriginal autochthonous relationships with land are distinct and quite unlike the distance relations of western representational systems' (2003: 293). We had to call out to the spirits of the dead for their assistance in obtaining food. Once we drifted out to sea with a broken outboard motor while it was getting dark and a rainstorm was coming. The grandchildren were instructed to call out for help to the spirit of their grandmother (deceased wife of Justin's elder brother) buried behind the mangroves bordering the coast, whereupon the outboard motor started working again and we got home safe. Another time, when with Justin and Joan and their children and families at Woolawunga, the rain (*pakadringa*) was driven away with a rite of throwing up sand as an inversion of the rain and an accompanying call ('puri puri tuwari'). Furthermore, Justin showed me the cavity at the beach in Imalu where a mythological rainbow serpent (*maritji*) dwelled. Only his people could approach the place unharmed, others he had to hold under his armpit. The smell from his sweat protected them. Further down the coast at a place called Teipu, characterised by a white cliff that formed an important source of clay for painting the body, the sea had taken a part of the land. The land that disappeared was experienced as such a loss that the people from the country Munupi decided to perform mortuary rites for it.

The nature of the close connections to the land suggest that the land is considered 'a living entity' (Grau 2005: 158). Our digging out of *kulama* yams for the annual increase ritual in April 1989, for example, would result in big rains in Justin's territory Rangini far away. Justin was convinced of this. He always spoke of the place as extremely rich in food. To feel well again he very much longed to go there after the death of his wife Joan. Mourning taboos made that he could not kill anything but, after initial cleansing rituals of the land had taken place, we spent two days clearing the road to Rangini. To be in his country in the company of his grandchildren, clearing it by burning pockets of tall grass, eased Justin's grief (cf. Ponsonnet 2014: 45). It also made him feel good to go with his children and grandchildren to all the places along the coast in Munupi closely associated with his beloved wife.

When the turtles vanished

Joan Puruntatameri was ill for a long time, at least for one and a half years. Several times she was on the verge of death, but every time she recovered. Joan stayed in Milikapiti on the north coast of Melville Island, where her eldest son lived and where she wanted to be buried next to her father. Her husband Justin and their children often went out hunting and gathering, believing it was the bush food that gave her strength. In spite of the fact that she preferred to stay home, Joan was sent to the hospital in Darwin on the mainland twice. The last time, the diagnosis had been clear. Joan was suffering from lung cancer: a big round spot, she related. The doctors could not do anything more for her.

As she lived on, Joan came to Pirlangimpi, on the sea strait between Melville and Bathurst Island in the northwest of Melville, where her husband, her five daughters and her youngest son lived. Her second eldest daughter, a health worker, and her youngest daughter, a teacher, took care of her on a daily basis. They were both single. Later they were assisted by a married sister. Joan's condition fluctuated. At times, she felt well. She played cards, danced and sang in the ceremonies, and even went out camping for a couple of days at Woolawunga, a northern beach location at Melville Island's west coast. At other times her condition worsened.

'They came say lovely sorry for me', said Joan, reflecting on the visits to her deathbed in the past few days. The senior Tiwi woman had recovered a bit and even taken a stroll to the beach. There, on the front beach, she was giving instructions for her funeral. She pointed to the waves. When the tide was out, she joked, she would have to be buried in the beach that had become exposed. Her foot would be revealed and stick above the sand when the tide came in, waving up and down with the incoming sea water. In other words, she would participate in the turtle dance, her foot making the char-acteristic movement with the flaps, at her own funeral. Joan happened to be the eldest person and leader of the turtle clan. Joan's clan used to perform the turtle and the tide dance.

On Tuesday 25 July 1989 she fell really ill again, leaving her bedridden. The next day her eldest daughter, on holiday in Perth, was called back. Things turn bad on Sunday 30 July. On the Monday afternoon Joan seems to be dying. In front of the house her husband, assisted by a son-in-law, has made a big shelter to provide shade with a roof of coconut palm leaves. Relatives from Milikapiti camp in the front yard, so do relatives from Bathurst Island who have come as well, as have almost all of Joan's children and grand-children. People from Pirlangimpi also come to visit. Justin sits at her bed. Joan sleeps most of the time. Now and then she opens her eyes, asking a female visitor where she had been all the time.

The following five days Joan remains severely ill. Twice her pulse fully dis-appears, but she comes to life again. Thursday 3 August, Joan's half-sister, a nun, arrives from Papua New Guinea. At noon the next day, women from Pirlangimpi say that Joan is 'nearly finished'. Her husband and children, and a lot of other people, in addition, are sitting inside with her. Most from Bathurst Island had gone back for payday. At the local health centre people from elsewhere call or have to be called by the daughters to keep them posted on Joan's condition. Her pulse falls away a few times. When I visit her in the late afternoon the next day, her hand feels cold. Joan has become very thin. Her eyes are half open and her mouth opens now and then too. A sarong has been draped over her lower body. Her half-sister and a female friend put pieces of cloth soaked in cold water on her side and chest. Joan wipes them away and scratches her shrivelled belly and breasts. Justin has come to sit with her too. Outside life goes on. Justin's siblings and their families camp under the shade at their respective fires.

That night, at a quarter past one, my Aboriginal neighbours wake me and say 'that old lady is finished'. We walk to Joan's house, talking about playing cards. The female neighbour is in an uplifted mood as she had been winning. It is relatively cold this night. About thirty metres from the house the neighbour lifts her arms and lets the sleeves of her sweater slide back a little: 'shaking', she says. Two metres towards the boundary of the front yard, she starts rubbing in her eyes, holding up her elbows to the same height. Next, she cries loudly, walks into the house, cries and returns wailing. She sits down with a group of women. The loud crying and wailing that comes from the house is the sound of the jungle fowl. All cry in the same way, go to the dying woman and come back. All the people who were inside leave the house. Joan has died. For ten days she has been severely ill. Trice her pulse disappeared. Justin is taken outside by two women, each holding him by one arm. They lead him from the stairs (to the house that is on little pilasters) to a chair. Justin sits down. Soon however he climbs up to the side of the little stairway (about 80 cm in height) and lets himself drop backwards on his back. He sits down on the chair again and sings. Two young men now hold his pulses. They walk with him to the side of the house, around the room with Joan's corpse, and return. Justin cries and sings.

Loud wailing and hitting oneself with the fists, or an object like a stick or a stone, is the conventional way of 'saying sorry' (*nuripmiori*). In saying sorry Tiwi address themselves towards the dying, the deceased, or the close relatives of the deceased. The dying, or when they are not able to do this, the close relatives of these persons or the dead stop the wailers, holding them so that they cannot harm themselves. Sometimes they say 'it is enough' (*weya tua*). These responses show that the wailing is accepted as an expression of relatedness and sorrow. In other words, the grief is shared and empathy shown by the self-wounding (Ponsonnet 2014: 45). Those who intervene, in turn, seek 'to protect the bereaved' and offer their 'support and concern' (Reid 1979: 330). The widower is looked after and people take care he does not inflict serious injuries on himself.

From 11 o'clock onwards people were to come. Joan's body had been laid in state on a steel frame of a bed under the shade in front of her house. Justin sits on the bed aside of Joan, his eldest sister on the other side of her. The daughters and sons are sitting close to their dead mother. A few relatives from Milikapiti do so as well. The division of people over space reflects social distance, the various categories of kin being grouped together and a gender division being upheld. The more distantly related people sit at some forty metres away from the corpse. On arrival, before they go to this place, they say sorry either facing the deceased or the bereaved. The visitors cry loudly. They might hit or harm a part of the body indicating their relationship to the dead (see further below). In saying sorry, men in general hit themselves with their fists on the loins and women do so with the right fist on the left shoulder (in the past, women carried a piece of bark to cover their pubic region in ceremonies, hence they could use only one fist).

A few metres from the bed an old man, Joan's son-in-law, is singing: '(dead woman saying:) I do not want all these *lorula* [semi-moiety, cluster of clans] to come up and dance for me'. Albeit Joan's daughters had been promised to this man, they did not marry him. Whenever he turned up, however, they had to provide him with food. The son-in-law lets his mother-in-law say that his people ought not to be obliged to attend and dance for her. Simultaneously, he refers to the customary avoidance relationship they had.

Then a first round of dancing starts. Two granddaughters of Joan are adorned with a goose-feather ball around their neck. The ball represents a baby (evoking the image of crossing the arms across the chest with the hands towards the shoulders as when a baby is carried on the shoulders) and elevates their status, that is, makes them 'important'. The emotions flare up when the coffin arrives and is placed in the shade by the ritual workers. Justin falls across the bed and tries to hit his head on the coffin. He is prevented from harming himself. A relative feels his head and looks through his hair to ascertain he is not wounded.

After the dancing a Roman Catholic priest celebrates mass. A choir from Bathurst Island sings the hymns in Tiwi. The audience assumes a rather passive mood until their emotions are suddenly triggered by a song about the deceased's country, Imalu. They wail and cry loudly, which intensifies when the lid is nailed on the coffin. This continues when the coffin is placed on the back of a truck for departure to the cemetery.

A few women are crying at a recent grave of a stillborn when the procession arrives. The workers scrape out the final bits of earth from the new grave that has been dug by them. Justin sings, hitting himself repeatedly with the fists at the loins. He makes the gesture of a butterfly, following his late (classificatory) father, and sings meanwhile. Next, he drops alongside the coffin. Together with two of his daughters he clings to the coffin, attempting to hold it back from being put into the grave. The coffin moves back and forth in the wrestle of the tangle of people. The deceased's daughters wail loudly. One who lived the last half year with her family at her mother's house has to be held back and seems on the verge of collapsing. Her tongue has turned blue, her belly is shaking fiercely, and she produces ripping, piercing wails. A granddaughter of the deceased, wearing a goose-feather ball, is in a similar state. The intense expression of emotion continues during the burial ceremony. Her father feels her forehead as she seems to faint from lack of oxygen.

The coffin is placed in the open grave. In the collective wailing the sound of the jungle fowl, the dreaming from Justin's people found in their country, can be recognised (cf. Basedow 1913: 310; Puruntatameri et al. 2001: 93). Justin stands at the edge and grabs a spade from one of the workers, who keeps holding the handle, and lets the blade land on his head several times. Next, he jumps into the grave and burns his pubic hair with a lighter. Meanwhile he sings: '(dead woman saying:) do not burn the whole lot (of pubic hair)/ because your wife might growl at you'. Thereafter, a dance and song ceremony around the closed grave commences. On this occasion not all of the relatives are allowed to dance yet. Those that are dance towards the grave. Joan's eldest son shows his emotion by performing an energetic dance, letting the sand spat spray up high. Two other men end their dance by making a somersault over the grave and landing on their back. Others throw the sand from the grave mound over their head. The ceremony concludes with Justin, the widower, having his whole body painted with white clay by Joan's half-sister. Meanwhile he performs a mourning song. He leaves his loin cloth, worn during ceremonies for years on end, on a stick at the grave. The children and grandchildren step over the grave mound before leaving the burial place.

On 10 August no one is on the beach at 5 pm. The far-out tide just starts coming in. A sea eagle is circling in the air. Justin walks onto the beach in a straight line from his eldest daughter's place towards his dinghy. Justin wears a loin cloth and a belt, his face and body still covered with white clay. He smokes. Justin tells that at Woolawunga and further north along the coast nothing is to be caught anymore. All fish and turtles have moved to Seagull Island (a sand island north off Cape van Diemen on the north-west coast of Melville Island, that is, north of Joan's territory Imalu). At the reefs near Woolawunga and further north there are no fish and turtles left. People have to wait for a fortnight before they can harpoon turtles again. Then the first turtle that will be caught cannot be cut before some totemic dances (turtle and tide) have been performed. Justin says that he cannot go in the dark to his place (because of his deceased wife's spirit) and, therefore, has to return home.

Ritual cleansing of the dead's territory

The night and early morning had been cold for Justin, who was wearing only a loin cloth. In the morning he had warmed himself at a fire. Later he sat in the sun to become warmer. Justin stuck to the rules he had set to himself as mourner. He was to stay away from places frequented by his deceased wife.

At 10 am the next day, 12 August, a cattle truck and four pickup trucks loaded with people leave Pirlangimpi for the dry-season camp at Woolawunga. The unsealed road ends at a high point. People split in groups, women and men separated, and members of the Puruntatameri and Tipaklippa patrilineages central. Justin is painted with white clay and wearing a new loin cloth and has a goose-feather ball around his neck. He sings and walks back and forth as a widower in ritual fashion: '(dead woman saying:) I am the head of the Rumakulumi [the people from Pirlangimpi]'. A classificatory mother of the deceased wails softly but audibly. The male workers, all in loincloths, collect branches with green leaves. Justin goes ahead, meanwhile singing: 'maybe that fire will go/they burn'em up to Tokwiangumpi (burial place of dead woman's father's brother)'. The men address the spirits of the dead with a prolonged mosquito call. This utterance is said to clear their voice. Justin sings: 'my friends (dead woman's fathers)/maybe there was talking about it all night'. Next the son of the dead woman's father's eldest brother's daughter holds his right hand up and calls out the place names given by Joan's father and his brothers and their father: 'Pulangumpura,

Watjutap, Putriamirra, Timilura, Purapuntri, Tokwiangumpi.' This means that these places are no longer taboo and the animals and fish can come back.

In an open space towards the beach the people of Joan's clan perform the turtle dance. Their leader composes the accompanying lyrics: 'turtle (*tarakulani*), we got to find it'. Then the group halts a second time, now on the edge of the bush and the beach, and once more does the turtle dance. Following this, they move on to the beach. The ritual drama that unfolds is the track of the turtle laying eggs on the beach. Where the farthest point of the track up onto the beach is located a woman and a man perform the turtle dance. They are paternal cousins of the deceased. At the spot besides where one normally finds the eggs Joan's granddaughters have to perform the turtle dance as well. Justin walks up to his ankles in the sea water and then up to this place on the beach. He sings: '(dead woman's father saying:) I am the big boss of this country.' The man who called out the place names does the same, but ritually cleanses his arms with sea water. Coming from the top and moving to the sea goes the main paternal cousin, a grandchild of Tipaklippa like Joan. Nearing the water, he makes the movements of the turtle as in the dance. Two steps into the sea he dives like a turtle and swims about ten metres under water. He resurfaces and hits first with his right arm and then with his left arm on the water. Finally, he ritually cleanses his whole body with water, removing the mourning taboos.

From the beach the people walk back into the bush to the place where Joan slept. There the ritual workers make a fire and ritually cleanse the place and the bereaved close relatives with smoke. Justin sings: '(dead woman saying:) I think he will put his footsteps on top of mine'. He hits himself and wails facing the spot where his deceased wife slept. The smoking is followed by a dance and song ceremony (*yoi*) in which the various categories of bereaved perform. Justin sings: '(dead woman saying:) they dancing where our country' and names all the places. Joan's paternal cousin (*mutuni*), and second in line of the turtle clan, holds a harpoon. He sings about the spiritual conception for the deceased's children (*mamurapi turah*), who are about to dance: 'He made a lovely two-sided barbed spear.' The dance enacts the spearing of the father by the spirit child in a dream. The singer cleverly relates that to the location, the deceased and her father Tipolei. The latter had a reputation as a fierce fighter and killer with barbed spears. Tipolei and his brother had blocked access to their land by placing upright several sets of two-sided, barbed spears cross-wise, and connected with a bush rope, on the beach of Woolawunga. Only after extensive negotiations, white people, who called Tipolei 'King', were granted access. Tipolei would have been the bereaved father (*unandani*), who 'made' the deceased, in the ceremony: the two-sided, barbed spear is female and can only refer to the dead daughter.

Other songs express how two other clans got frightened, to which a ritual worker added: 'He [Tipolei] made the fire all around, they run away.' The maternal children (*mamurapi pulanga*) conventionally call out to be breastfed to no avail. Hence the accompanying song to their dance: 'Oh, you got no milk/so we got to swim in the sea (make them cold).' The lyrics make clear they are members of the turtle clan. Two sons-in-law dance with small branches with leaves wiping their shoulders, concluding with putting these aside. The dance (*impala*) for the deceased mother-in-law (*amprinua*) emphasises the avoidance relationship; the shoulder being the body part where one feels pain when something is wrong with the mother-in-law. Instead of mentioning the hurt shoulder directly, the lyrics often tell of a wing or a branch broken off. In this case they allude to that image: 'They carry wood and fell down with that wood.' Finally, the ritual workers (distant in-laws, *ambaruwi*) and the widower (*ambaru*) dance. Justin spirals around his lined-up daughters with both hands in the air. The workers dance in their distinctive styles. Joan's maternal half-sister (*putakka*) is placed in the middle. They kick her leg, the body part hurt or 'lost' due to the death. Justin, the widower, also spirals around her, hands up and opening and closing his legs.

Next, he lies down on his belly on the spot where his deceased wife had slept. Justin cries. Then he turns on his side and sings: '(dead woman saying:) maybe we will have sex while I am sleeping.' He rolls in the hot sand and ashes from the burnt down fire. Justin hits himself with a piece of wood. He wails and sings: 'She used to live with me at Pirlangimpi, but she left me there.' Meanwhile another senior man and relative takes away the piece of wood from the widower. This gesture evidences protection and concern. Comfort, according to Berndt, is also obtained from talking with the dead person in the lyrics: 'The conversational dialogue, reminding the spirits of actual incidents in the past and pretending to plan for the future, serves to provide some comfort for the mourner, in an apparently casual fashion which dispenses with most of the more overt expressions of grief' (Berndt 1950: 305). The widower continues this dialogue at Putramirra, a place and beach further north where the people attending the ceremony move on to next. The main reason is to make the calls of place names, indicating that the area is no longer taboo (*pukamani*) for turtles, fish, wallabies and so forth. The man who made the calls with his right hand raised had been asked to go up front at Woolawunga and Putramirra because his own mother, a paternal 'sister' (niece) of the deceased, had been buried there. Her spirit was considered powerful in setting the area free for the animals and for hunting and gathering.

The following afternoon Justin's youngest son and a friend went to the reef in front of Woolawunga. They harpooned a turtle. This turtle was brought to the front beach at Pirlangimpi, where Joan's paternal cousin and a few others of the turtle dreaming danced before it could be cut. This had to be done with the first turtle after it had been taboo. One of the deceased's daughters stated: 'All turtles are back now, but last week nothing at Woolawunga, no turtles and no eggs.' Her father, however, told me he followed his father and father's father. That is to say, he stuck to mourning taboos. He could not eat food – turtle, fish and so forth – from Woolawunga and that area, and also no animals such as bandicoot and wallaby. It was alright to eat them for other people, but Justin had to wait until after the wet season, when the rain had cleansed or washed away the *pukumani* (taboo) from the land.

Postfuneral rites from Pirlangimpi to Karumurarimili

The widower would have a first, partial release from his mourning taboos on the first of October, at the conclusion of the final rites for his deceased wife, when I would give him a ritual washing (*moluki*). Joan's half-sister, who lived in Port Moresby, had asked Justin to hurry up with the postfuneral rites so she could still attend them. Although he would have had to wait for at least half a year, Justin had said 'alright'. The postfuneral rites (called *pukamani*) for Joan would start in Pirlangimpi but end in another part of Melville Island, on the hill near Milikapiti, where Joan's father and brother were buried. The burial place is called Karumurarimili or Karangumungumili.

Justin had his face painted with yellow and white stripes. He had picked up the yellow ochre on the way back from Milikapiti, where Joan's half-sister had wanted to see her relatives. It is no coincidence that he had chosen these colours as they are emblematic for the deceased's (and his) country: the superb white clay is from Teipu and the best yellow ochre comes from Imalu, respectively south and north of Woolawunga, in the country Munupi (the best red ochre comes from somewhere else, namely Arapi in the southwest of Bathurst Island). Another association with the land concerns the pubic hair Justin scorched off in Joan's grave. The bodily hair is equated with the long grass that is burned down in the dry season (Venbrux 1995: 143). Malinowski already notes that among Aborigines the marital bond tends to last after death, 'and expiation must be made for the eventual new union' (1913: 87). Justin explained that he had to wait for a year before a promised wife could join him, another year for the second, and one more for the third. The regrown pubic hair was associated with the tall grass where one could have a tryst with a new lover.

In the morning of 18 August Joan's house is cleansed with smoke. Although Justin had his own little house, which he reoccupied the day before, in another part of the township, the encounter with the place of her death, taboo for the intermediate period, was very emotional. He sang intimate mourning songs and had to be restrained by others from harming himself. Next, we went to a burial place in the bush in the southwest of Melville Island for the final mortuary rites for a homicide victim that took one and a half days and the night in between (see Venbrux 1995: 183–222). Striking in the many death rites that Justin and his children and grandchildren attended was their sharing of grief with the newly bereaved, or with the previously bereaved in the case of postfuneral rituals, and obtaining support from one another. This was the case at a funeral of a man at Wurrumiyanga (Nguiu) on Bathurst Island on 24 August, the final rites for two women there on 27 August, intermediary rites for a man in Pirlangimpi, Melville Island, on 28 August, and the final rites for a man there on 2 September, where Justin bit his goose-feather ball in his dance performance, and a funeral of a woman in Milikapiti, Melville Island, on 9 September 1989. Besides obtaining solace, Joan's close relatives mobilised support for the final rites for her. The day after we had cut free the road to his (sub)country Rangini, Justin felt emotionally ready to commence the *pukumani* for his deceased wife.

Shortly after midday, on 11 September, people came together under the shade, covered with coconut palm leaves, in front of Joan's house in Pirlangimpi. The men are wearing loin cloths. Many, including Justin's daughters, are painted with white clay. The youngest daughter was teaching in Bathurst Island, but had returned home for the rites. Justin comes from the house of another daughter and enters Joan's house. He visits the room where his beloved wife had died. Joan's half-sister, who stood at the bed, told me Justin had thrown himself on top of Joan, saying: 'she was pretty'. The woman had tried to hold him back because her sister was still breathing. At the time of the ritual cleansing of the house Justin had prostrated himself on the bed again. Meanwhile he performed a mourning song. This time he wails loudly. He leaves the house, keeps standing in front of it and sings again a widower song, which he repeats under the shade. Ponsonnet (2014: 45–46, 205), writing on the Dalobon Aborigines, sees the visit to a place associated with the dead person as a strategy to overcome grief: 'Grief is intentionally revived on this occasion, but strictly monitored (one makes sure that close relatives will be around providing support, etc.)' (ibid.: 46). Justin's utterance at the time of Joan's death referred to her beauty. Joan was particularly proud of a photograph of them as a young couple (Elkin 1964: Fig. 8, opposite p. 95). She told me once, when we looked at the photo, that she was admired for her protruding breasts. Her absence now must have intensified rather than assuaged his grief. Justin put two axes ready under the shade and gave the sign that the ceremony could start.

The oldest grandson from Tikaklippa, the deceased's grandfather, composed the first song and danced as if cutting a tree. The various categories of bereaved dance as well, the women and men always as separate groups. The purpose of the axe-giving ritual (*walemani*) is to commission the ritual workers who will make the carved and painted grave posts to be erected around the grave at the conclusion of the final rites. At the end of the dance the axes are handed to selected workers. A second small ritual concerns the presentation of fire (*ikwonni*) to dry and darken the debarked tree trunks. On this occasion matches changed hands, whereas the dance gestures showed the old way of making fire with sticks. Finally, the paints (*tilamara*) to put on the carved posts are given. In this third ritual the gesture was made with a piece of white clay. The accompanying dance of painting one's face and arms underscores the close association between body and tree or land. During the three rites the ritual workers dance in the background.

At the conclusion the two axes are placed in the middle of the dancing ground. Justin, the actual widower, dances around with his hands in the air and opening and closing his legs towards the axes. He is the last to leave the place, after having performed a mourning song. He expected that eleven grave posts from Milikapiti and eight from Pirlangimpi would be made for his deceased wife.

The next morning Justin pointed out a bloodwood tree that had to be cut at Blue Waters, an ancestral burial place not far from Pirlangimpi in Joan's country. He had drawn the desired shape of the commissioned grave post on the back of a box of matches. The grave posts can be seen as expressions of sorrow (Hart 1932: 18), yet as aesthetically pleasing tangible and solid bodies, reminiscent of the deceased in shape and by means of the painted design and colour, they also offer solace (Venbrux 2017b). Their proneness as objects of consolation is further due to their link to the deceased's country, like in the past blood had to drop on the grave when people cut their heads when grieving, a blood-like substance flows from the tree onto the land when cut. The treatment of the tree trunks, of a most vital type of eucalypt, resembles the ritual processing of the close relatives of the deceased, including cleansing with smoke and lying down (associated with sleep and death), when painted and subsequently standing up (associated with being alive) around the grave (Venbrux 1995: 193–194). The same day that the grave posts are cut Justin departs to Milikapiti because his eldest sister is dying.

The area where Joan is buried is still taboo (*pukamani*). On 17 September a man returning to Pirlangimpi from taking a couple going hunting and gathering rolls over with his car on the dirt road leading to the burial place. The driver dies later in the day. The next day I meet Justin at the funeral of his eldest sister (in which I act as ritual worker) in Milikapiti. He enquires about the fatal accident. Justin is adamant that they should have waited till after the ceremony for his deceased wife. It was too dangerous, because her father was there now. Joan's father, named Plakwuri (also known as Tipolei), was a 'killerman' (*kwampini*).

In Pirlangimpi a fight had erupted over where the man killed in the accident had to be buried. Two powerful half-brothers from Bathurst Island wanted to have him buried there with their fathers. Initially they persuaded the reluctant widow, but all local close relatives turned against it, including another half-brother and the man's biological father. The latter, painted all in white, hit himself on the head facing the Bathurst Island leader, who walked away in anger from the inquest, stating that if he was buried on Melville Island they would not attend the funeral. Both gestures signalled a breaking of ties within the family over the all-important issue of the place of burial. At stake was the emotional attachment and rights to land, and hence the spirit-promoted well-being, for the patrilineal descendants. Amidst tension the corpse spent a night in jail, locked up, at the local police station in Pirlangimpi. On 20 September the funeral takes place and the man is buried not far from Joan. Justin performs a mourning song as a widower when he arrives at the burial place and walks towards Joan's grave. He walks back and forth in widower style, but mainly at the head end of the grave, where his loin cloth left behind is still attached to a stick. He clears the grass away from the grave (cf. Spencer 1914: 232, 234). After the burial ceremony his daughter is crying at the grave together with a relative who had not been able to attend Joan's funeral. Justin is in conversation with his wife's spirit in an *ambaru* song. Three days later the funeral of Joan's son-in-law, who never got her daughters promised to him, takes place at Pawularitarra in the southwest of Melville Island. (He is to be buried next to the homicide victim as they have one grandfather in common.) The widow wants Justin's daughters to dance around the coffin as she did. Justin has to dance them towards the grave later on after the burial and composes the accompanying song. He then leaves with his son and daughters, saying to a senior man from Milikapiti: 'you take over now'.

The first of the intermediary rites (*ilaningha*) for Joan is at the place where she was laid in state, under the shade in front of her house, on 25 September. Justin wears a goose-feather ball. He and his daughters are painted up. The participants perform the turtle dance. The sons of Joan's eldest clan brother have to throw up the goose-feather ball three times. Having performed this action, they are considered the new initiands, that is, the clan's new generation to carry on the cultural practices. The ritual workers and the widower end the dancing with their specific performances. The people disperse, but Justin stays behind. He performs a mourning song while standing in front of the house, looking through the louvers of the window on Joan's empty bed.

Midday, 28 September, Justin wearing a loin cloth, face and body painted, and with a goose-feather ball around the neck, walks in a straight line to Joan's house where further intermediary rites are about to start. The ritual workers rake the sand in the shade and remove stones, twigs and so forth, so that none of the dancers will hurt their feet. Many come to this *ilaningha* in which the dreaming of the deceased stands central. It means that the turtle dance is performed by the various categories of bereaved, with the exception of the workers and the widower, while they themselves may have another dreaming. Important in this *ilaningha* 'right through' is an energetic dance in which men throw up dirt with their feet (*ampikatoa*). The insertion of either fast and energetic or slow dances to respectively heat up and 'cool down' (Seligman 1932: 198n1) the emotions in Tiwi mortuary rites is part of a 'regulation of grief' (Jedan, ch. 1 in this volume) that offers consolation. The intermediary rites would go on forever unless one of the chief mourners calls for the 'breaking' of their performance and moving on to the concluding rites or *iloti* (meaning: 'for good'). To this end the rite in question is also carried out at the final destination, crossing in this instance a distance by road of about 80 km.

Justin departs for Milikapiti to keep an eye on the preparations for the final rites. These commence at 5 pm on 30 September. The location is the burial place on the hill where Joan's father and brother have been buried. A new grave has been made with some of her clothes and personal belongings (these items can substitute for a corpse, see Venbrux in press) close to the grave of her father Tipolei with old and weathered posts. At this stage, people have to stay some 100 metres away from the new grave, marked off with a display of silkscreen printed fabrics in Tiwi design. A large shade covered with coconut palm leaves has been erected alongside the grave. At the grave is a mast with a flag, also of the costly Tiwi design. Justin addresses the grave and performs a mourning song. He sings: 'my wife, your mother, she is glad/because we meet together'. The widower continues, but now singing with his wife's voice: '(dead woman saying:) so you got to grab me tonight' and '(dead woman saying:) you should take your clothes off'.

Then a ritual cleansing with smoke takes place. The bereaved's anger about the death is conventionally directed at the cultural hero Purukupali, who during creation brought death into the world. Justin sings: 'that Purukupali is rubbish/he was telling people we got all to die'.

He proceeds: 'that man Purukupali, he was saying we all die'; 'spirit one, they should go chasing him away'. The ritual workers then go ahead towards the new grave posts lined up between the shade and the newly constructed grave. They call out and chase the spirits of the dead away with sticks. All the participants cry and wail loudly at the posts, holding these carved and painted sculptures. The encounter triggers strong emotions.

As always people are supposed to listen to the lyrics of the songs first and then cry. Justin's elder paternal half-brother sings: '(dead woman saying:) you didn't say hello to me and your brother/and we got a lot of kids', and then he wails. A classificatory mother of the deceased: '(dead woman saying:) my father, I call his name/I am not frightened by the people from Imalu [his country]/but got that lady from you.' The widower's next song is of a more intimate nature: '(dead woman saying) you should take all the clothes off/come straight to me with no clothes'. He cries and tells we will camp and sleep here tonight. Justin proceeds: '(dead woman saying:) he is coming/and he will take me (have sex with me) this night'; '(dead woman saying:) now he is feeling shaking for me/but we got to go in somewhere/because he is feeling shaky (aroused)'; '(dead woman telling her children:) you fellows sleep far away/your father gonna sleep with me/because for three weeks we did not sleep together'.

Time and again the words are followed by expressions of sorrow and crying by the performer. The widower's elder brother sings: '(dead woman saying:) you are the one now/I was looking for you/you are a good man/you got to look after my kids'. Justin points into the distance in the open forest where it looks as if two fires are lighting up. He tells me the spirits of the dead camp there. We feel a cold breeze. The spirit of Joan's father has arrived, Justin says. Justin's paternal half-brother composes the following song: 'those three men (Joan's father and his two brothers: Tipolei, Murantumolia and Marapauma) whisper/we sleep close to them/we got to kill these people with a spear/they tell their wives not to lit the fire/because they (the prospective victims) might get fright'. Justin tells me we can now open up the road where the car accident occurred near Pirlangimpi. It requires a cleansing ritual with smoke to remove the taboo from the place. 'No more spirit there again', he says. We sleep and the fires burn down. I wake to Justin's singing: '(dead woman telling him:) kill the fire/you and me, we got to sleep in the darkness'; (dead woman saying:) first we got to it once/and then take all the clothes off/ and sleep naked'. Before daybreak we hear the birds.

Justin's brothers call out. The one sings about the three brothers again, saying: 'He is fine cheek to kill (hit) her house'; the other one sings they hate his voice calling out, referring to his dreaming, the cockatoo. Justin relates in his song that someone threw a spear; 'all grannies got up and asked: "who did that?"/and some said: "oh, his wife, there was talking about her"'. He alludes to a historical song and at the same time to the first night of the annual yam ritual, the night of sorrow, when the participants lay down and sing about their grief and grievances, voicing trouble and argument. He later pursues it in the lead up to the *iloti*: '(dead woman saying:) maybe our father got to have *ajipa* (the highlight, with polychrome body paint and the most clever and intricate songs, of the yam ritual) today/my husband he is finished/ we will be finished after that'. Justin continues: 'she is still underneath the fire (in the ceremonial earth-oven like the yams) at her father's burial place Karumurarimili/a big crowd has come (to listen)/we finish today.'

In a different style of mourning song (*mamanakuni*) a bereaved classificatory mother includes Joan's brother (who died in 1984) buried nearby: '(dead man telling his deceased sister:) sister (waving her), she is here, our mother/ she is here with milk, sitting down/she got drink for us'. The *ambaruwi* (in-laws and ritual workers) sing about 'a good hiding' and 'jealousy'. The actual widower sings: '(dead woman saying:) I have a spear from Tipolei/made red in front/I am holding the spear/and say: "my husband and I sleep here!"'

For the people who arrive first in the morning there is another cleansing ritual with smoke. Next the ritualists go to the dancing ground under the shade. One man goes ahead, like he did at Woolawunga beach, calling out all the place names from the deceased's country. Then the *ilaningha*, performed earlier in Pirlangimpi and now carried over in space, is enacted. Joan's dreaming, the turtle, stands central in dance and song. A senior man initiated the first song: 'she is a female turtle with a big name (meaning, she is important)/when she comes out of the sea/she spits water up:/pep-pep-pep' (another way of saying, 'you fellow all come good out, when you are a good dancer'). This accompanied the turtle dance. He continued: 'where the sand bank is, they got that turtle'. The oldest grandson from Tikaklippa, the deceased's grandfather, joined in: 'she is safe at Rangmunarini (near Salt Creek in her country)'; 'she run away/long way'; 'big tide is taking her'; 'big wave (middle of the water)/and she is on top of the wave'. This intermediary rite could go on and on as long as no one called for the final rites, *iloti* (meaning: 'for good'). Justin concluded with marking a classificatory father by making the actions of a butterfly and calling himself a butterfly.

The iloti is comprised of dances and songs marking dreamings and the various categories of bereaved (for a detailed discussion, see Venbrux 1995). The spatial dimension came most to the fore in the performances of the cousins with a common paternal grandparent (*mutuni*), maternal (classificatory) siblings (*paputawi*) and the actual widower (*ambaru*). The first mentioned lost half of their face, which was marked by having it painted only half or in two halves of different colour, and by slapping the face, either by oneself or being hit by others. Tipaklippa's oldest grandson, who also performed the ritual cleansing of Joan's country so that turtles could be caught again, sang: 'I climbed that hill and was on top'; 'big waves at Imalu (the deceased's country) came/and wiped my face'; 'I had a sore face, all the way from my country'; 'that is a boil I had on one side of my face'; 'at Tupulurupi (grandparental burial place) they gave me sorry/the earth was shaking/they said: "sorry for you, you got one side boil"/and they fix my face'. Another one sang: 'I had a sore face, all the way from my country'; 'here at Karangumungumili (Tipolei's grave) my face fell down'. The *paputawi* lose their leg, that is, the deceased sibling, and have the leg marked with paint and hit. The accompanying song to the dance: 'I came down from a bumpy hill/and I broke my leg'. Justin, the widower, combines the marking of his relationship to the deceased and his dreaming: 'I am naughty'; 'a jungle fowl calling out/at that place where I take my wife'; 'kurupu-kurupu (calling out where a lot of mess)/at that place to take her'; 'the bird makes an action with his eye'.

During a break in the dancing people leaned on the colourful carved posts lined up at the grave side of the shade. For Spencer 'there was nothing quite so picturesque' as a ceremony like the one he 'saw on Melville Island' (1914: 239). Goodale (1971: 331) was overcome by the feeling that everything was 'alright' in the world. The mortuary rites are both aesthetically and intellectually challenging, with many subtle clues concerning connections and identifications (Venbrux 2009, 2017b; cf. Turpin and Fabb 2017). As Kertzer (1988: 101) puts it, ritual 'creates an emotional state that makes the message uncontestable because it is framed in such a way as to be seen inherent in the way things are. It represents a picture of the world that is so emotionally compelling that it is beyond debate.' It is not so difficult to imagine that something so beautiful offers some consolation. The *ambaruwi* put Joan's half-sister central in their final dance, the striking family resemblance triggers emotions. Towards the conclusion of the ceremony the workers lift the grave posts from where they stand and place them around the grave.

In paying the ritual workers yellow ochre is mentioned rather than the money that changes hands. The yellow ochre directly refers to Joan's land. The last act is the ritual cleansing of the widower with water. While the water is poured over him and washes off his body paint, Justin sings: 'Big waves from the sea at Imalu (Joan's country) wash my body off'/'big waves, and I swim without clothes, and those waves smack me (my private parts)'. A paternal cousin of his has the last word: 'you should take all your clothes off/ let your clothes fall off/let everybody see you/what you got!' The attribution of jealousy to the spirits of the dead indirectly indicates that the living are better off. What is more, the deceased has been given all her dues. In the last ritual act most bereaved people, according to Boston and Trezise (1987: 99), 'wish to do what they feel the dead person would have liked' as this feels right and gives consolation.

Conclusion

In Tiwi society the land as 'consolationscape' is integrated with a person's worldview, one's country being a prime source of well-being and identity. In this chapter I followed what happens in terms of grief and solace when a senior Tiwi person dies. As the ritual drama unfolds, the bereaved are offered support and protection, the sharing of pain and grief, and a management of the emotions and timing of the rites on a needs basis. The spirit of the deceased is sent and guided on a spatial trajectory, enabling a gradual transformation of the ties between the living and the dead, towards a spiritual dwelling in the country and hence reinvigorating the ancestral powers that sustain the life of its people. The performances in multiple media, including song, dance and the visual arts, move the bereaved to change perspective and therefore ease the grief.

References

Basedow, H. 1913. Notes on the natives of Bathurst Island, north Australia. *Journal of the Royal Anthropological Institute of Great Britain and Ireland* 43, 291–333.

Berndt, C. H. 1950. Expressions of grief among Aboriginal women. *Oceania* 20/4, 286–332.

Bloch, M. 1988. Death and the concept of person, in *On the Meaning of Death: Essays on Mortuary Rituals and Eschatological Beliefs*, edited by C. Cederroth, C. Corlin and J. Lindström. Uppsala: Almqvist & Wiksell, 11–29.

Boston, S. and Trezise, R. 1987. *Merely Mortal: Coping with Dying, Death and Bereavement*. London: Methuen.

Brandl, M. 1971. *Pukamani: The Social Context of Bereavement in a North Australian Aboriginal Tribe*. Unpublished PhD Thesis, University of Western Australia, Perth.

Cox, L. 2010. Queensland Aborigines, multiple realities and the social sources of suffering, part 2: suicide, spirits and symbolism. *Oceania* 80/3, 242–262.

Durkheim, E. 1912/1995. *The Elementary Forms of Religious Life*. New York: The Free Press.

Elkin, A. P. 1964. *The Australian Aborigines*. New York: Doubleday.

Glaskin, K., Tonkinson, M., Musharbash, Y. and Burbank, V., eds. 2008. *Mortality, Mourning and Mortuary Practices in Indigenous Australia*. Farnham: Ashgate.

Goodale, J. C. 1971. *Tiwi Wives: A Study of the Women of Melville Island, North Australia*. Seattle: University of Washington Press.

Grau, A. 1983. *Dreaming, Dancing, Kinship: The Study of Yoi, the Dance of the Tiwi of Melville and Bathurst Islands, North Australia*. Unpublished PhD Thesis, The Queen's University, Belfast.

Grau, A. 2005. When the landscape becomes flesh: an investigation into body boundaries with special reference to Tiwi dance and Western classical ballet. *Body & Society* 11/4, 141–163.

Hart, C. W. M. 1928–29. *Munoopula*. Unpublished geneaologies.

Hart, C. W. M. 1930a. The Tiwi of Melville and Bathurst Islands. *Oceania* 1/2, 167–180.

Hart, C. W. M. 1930b. Personal names among the Tiwi. *Oceania* 1/3, 280–290.

Hart, C. W. M. 1932. Grave posts of Melville Island. *Man* 11–12, 18.

Hart, C. W. M. and Pilling, A. R. 1960. *The Tiwi of North Australia*. New York: Holt, Rinehart and Winston.

Hoy, W. E., Mott, S. A. and McLeod, B. J. 2017. Transformation of mortality in a remote Australian Aboriginal community: a retrospective observational study. *BMJ Open* 7: e016094. doi:10.1136/ bmjopen-2017–016094

Kertzer, D. I. 1988. *Ritual, Politics, and Power*. New Haven, CT: Yale University Press.

Klass, D. 2013. Sorrow and solace: neglected areas in bereavement research. *Death Studies* 37/7, 597–616.

Klass, D. 2014. Grief, consolation, and religions: a conceptual framework. *Omega: Journal for Death and Dying* 69/1, 1–18.

Langton, M. 2002. The edge of the sacred, the edge of death: sensual inscriptions, in *Inscribed Landscapes: Marking and Making Place*, edited by B. David and M. Wilson. Honolulu: Hawai'i University Press, 253–269.

Maddrell, A. and Sidaway, J. D. 2010. Introduction: bringing a spatial lens to death, dying, mourning and remembrance, in *Deathscapes: Spaces for Death, Dying, Mourning and Remembrance*, edited by A. Maddrell and J. D. Sidaway. Farnham: Ashgate, 1–18.

Malinowski, B. 1913. *The Family among the Australian Aborigines: A Sociological Study*. London: University of London Press.

Metcalf, P. and Huntington, R. 1991. *Celebrations of Death: The Anthropology of Mortuary Ritual*. Cambridge: Cambridge University Press.

Morphy, H. 1984. *Journey to the Crocodile's Nest. An Accompanying Monograph to the Film Madarrpa Funeral at Gurka'wuy*. Canberra: Australian Institute of Aboriginal Studies.

Morphy, H. 1995. Landscape and the reproduction of the ancestral past, in *The Anthropology of Landscape: Perspectives on Place and Space*, edited by E. Hirsch and M. O'Hanlon. Oxford: Clarendon Press, 184–209.

Mountford, C. 1958. *The Tiwi, Their Art and Ceremony*. London: Phoenix House.

Myers, F. R. 1986. *Pintubi Country, Pintubi Self: Sentiment, Place, and Politics among Western Desert Aborigines*. Washington, DC: Smithsonian Institution Press.

Myers, F. R. 2002. Ways of place-making. *La Ricerca Folklorica* 45, 101–119.

Norberg, A., Bergsten, M. and Lundman, B. 2001. A Model of Consolation. *Nursing Ethics*, 8/6, 544–553.

Osborne, C. R. 1974. *The Tiwi Language*. Canberra: Australian Institute of Aboriginal Studies.

Pilling, A. R. 1958. *Law and Feud in an Aboriginal Society of North Australia*. Unpublished PhD thesis, University of California, Berkeley.

Ponsonnet, M. 2014. *The Language of Emotions: The Case of Dalabon (Australia)*. Amsterdam: John Benjamins.

Puruntatameri, J. et al. 2001. *Tiwi Plants and Animals: Aboriginal Flora and Fauna Knowledge from Bathurst and Melville Islands, Northern Australia*. Darwin: Parks and Wildlife Commission of the Northern Territory/Tiwi Land Council.

Reid, J. 1979. A time to live, a time to grieve: patterns and processes of mourning among the Yolngu of Australia. *Culture, Medicine and Psychiatry* 3/4, 319–346.

Robinson, G. W. 1990. Separation, retaliation and suicide: mourning and the conflicts of young Tiwi men. *Oceania* 60/3, 161–178.

Rose, D. B. 1992. *Dingo Makes Us Human: Life and Land in an Australian Aboriginal Culture*. Cambridge: Cambridge University Press.

Rose, D. B. 1996. *Nourishing Terrains: Australian Aboriginal Views of Landscape and Wilderness*. Canberra: Australian Heritage Commission.

Scott-Clark, C. and Levy, A. 2006. The land of the dead: how did two tiny islands off Australia's coast come to have the highest suicide rate in the world? *The Guardian*, 24 June 2006. Online: https://www.theguardian.com/world/2006/jun/24/australia.adrianlevy

Seligman, C. G. 1932. Anthropological perspective and psychological theory. *Journal of the Royal Anthropological Institute of Great Britain and Ireland* 62, 193–228.

Spencer, W. B. 1914. *The Native Tribes of the Northern Territory*. London: Macmillan.

Stanner, W. E. H. 1979. The Dreaming, in: *White Man Got No Dreaming: Essays 1938–1973*. Canberra: Australian National University Press, 23–40.

Swain, T. 1993. *A Place for Strangers: Towards a History of Australian Aboriginal Being*. Cambridge: Cambridge University Press.

Tonkinson, M. 2008. Solidarity in shared loss: death-related observances among the Martu of the Western Desert, in *Mortality, Mourning and Mortuary Practices in Indigenous Australia*, edited by K. Glaskin, M. Tonkinson, Y. Musharbash and V. Burbank. Farham: Ashgate, 37–53.

Turpin, M. and Fabb, N. 2017. Brilliance as cognitive complexity in Aboriginal Australia. *Oceania* 87/2, 209–230.

Van Gennep, A. 1909/1960. *Rites of Passage*. Chicago: University of Chicago Press.

Venbrux, E. 1993. Les politiques de l'émotion dans le rituel funéraire des Tiwi d'Australie. *L'Ethnographie* 89, 61–77.

Venbrux, E. 1995. *A Death in the Tiwi Islands: Conflict, Ritual and Social Life in an Australian Aboriginal Community*. Cambridge: Cambridge University Press.

Venbrux, E. 2009. Social life and the dreamtime: clues to creation myths as rhetorical devices in Tiwi mortuary ritual. *Religion and the Arts* 13/4, 464–476.

Venbrux, E. 2015. 'White people, or might be devils': commemoration of the 1705 Dutch landing at Piramparnalli, Melville Island, North Australia, in *What's Left Behind: The Lieux Mémoire of Europe beyond Europe*, edited by M. Derks, M. Eickhoff, R. Ensel and F. Meens. Nijmegen: Vantilt, 126–132.

Venbrux, E. 2017a. How the Tiwi construct the deceased's postself in mortuary ritual. *Anthropological Forum* 27/1, 46–62.

Venbrux, E. 2017b. L'art, la religion et la mort dans les îles Tiwi, in *L'effet boomerang: Les arts aborigènes et insulaires d'Australie*, edited by R. Colombo Dougoud. Genève: Musée d'ethnographie de Genève (MEG), 97–107.

Venbrux, E. in press. Destroying objects, keeping memories, in *The Materiality of Mourning: Cross-Disciplinary Perspectives*, edited by Z. Newby and R. Toulson. London: Routledge.

Watson, C. 2003. *Piercing the Ground: Balgo Women's Image Making and Relationship to Country*. Freemantle: Freemantle Arts Centre Press.

8 Rituals, healing and consolation in post-conflict environments

The case of the Matabeleland Massacre in Zimbabwe

Joram Tarusarira

Introduction

Death, when natural or expected, is one demonstration of the paradoxes of life. While it is a phase of life, it is a phenomenon with which people across the world struggle to come to terms. Thus, in the Shona language spoken by eight out of ten Zimbabweans, there is a saying – *rufu har-ujairike* – which means that death is something to which one can never become accustomed. Regardless of how often one encounters death, the feelings of sadness, emptiness, trauma, helplessness, brokenness, wounded-ness, and pain are always newly experienced. Another person's death interferes with an individual's psychological and physical stamina. It dislocates the family's integrity by severing one of its members (Kumbirai 1977: 123). It is an unsettling, threatening, shocking, dreaded, disrupting and suspicious phenomenon (Kaguda 2012). The intensity of death's negative psychological and physical effects is accentuated when it is unexpected, or when its cause appears shrouded in mystery. A loved one's violent and humiliating death inflicts a mental scar. Both scenarios – deaths shrouded in mystery and violent deaths – are the focus of this chapter. Human beings try to find ways of dealing with bereavement, reducing the distress produced by the experience of death, restoring as much as possible normal functioning, and fostering resilience against past as well as unspecified future losses (see Jedan, ch. 1 in this volume). Healing and consolation are thus ways of addressing the dislocation caused by death. Various strategies, ranging from prayer, religion, and spirituality to ritual practices, have been deployed to deal with the legacies of death and loss. This chapter discusses how the people of the Matabeleland and Midlands provinces in Western Zimbabwe have dealt with the legacies of the state massacre called *Gukurahundi*, a Shona word for 'the rains that wash away the chaff'. The chapter addresses the role of rituals associated with the dead among the Ndebele as pathways towards healing and consolation, although it makes occasional references to the Shona culture, which is also found in Zimbabwe and which has cultural commonalities with the Ndebele culture. The case illuminates our understanding of how traditional methods are deployed to deal with dying, death, mourning and remembrance in pursuit of healing and consolation. It is argued that religious and cultural practices, which are the *doxa* of a people, resurface in the face of existential questions such as death.

Death wounds

Consolation and healing are engendered by woundedness, which can take various dimensions. The obvious wound is the one created by the absence of someone who used to be in the family. The fact that the descendants will never see them again is traumatising in itself. Woundedness can refer to brutal harm – i.e. the physical, psychic, economic or emotional harm to the victim's personhood. After the violent death of a family member or friend, the survivors have to deal with the real or imagined injury to the deceased. In cases where the survivors have witnessed the death, every moment of remembering the occasion evokes negative painful emotions. Woundedness can be based on ignorance regarding the circumstances surrounding a loved one's death. This wound is willingly sustained by the perpetrators of violence when survivors are confronted with questions such as 'Who was behind the gun?' or 'How was my daughter or son abducted?' Not knowing the truth creates a wound within the bereaved. As in the Matabeleland massacre, when deaths were politically motivated, some deaths are shrouded in mystery. People 'disappeared' and were killed in secrecy, only to be discovered as decomposed, abandoned bodies, if they were discovered at all. Such abandoned bodies are often found with scars and bruises which speak only of how brutally they were violated. This, therefore, raises many uncomfortable and unsettling questions. Ignorance of how a family member or friend died is in itself torture for the survivors. However, it is contestable whether truth-seeking brings about healing or catharsis (Hayner 2011: 151). Woundedness can emanate from how the dead person was denied justice as a member of the political order from which they were entitled to human rights. This is interpreted as social death or political death. To imagine that the victim was denied social and political life, while the perpetrators continue to enjoy life socially and politically, is rubbing salt into a wound. It is painful to see the dead being casually treated in the service of political ideology without remorse on the part of the perpetrators. To make matters worse, perpetrators deny responsibility for the wounds they inflict. If they speak of what they have done, their words will only be in self-defence and justification, causing further pain to the survivors. Not only are the perpetrators unaccountable for the wounds they inflict, but they also make no effort to confess, apologise, atone for or make amends for the perpetration of political violence (Philpott 2006: 16–18). These dimensions of woundedness together explain why consolation and healing are a priority in post-conflict scenarios, especially where deaths are involved.

The case of *Gukurahundi*

Former President Robert Mugabe and his party, the Zimbabwe African National Union–Patriotic Front (ZANU PF) – unleashed an operation known as *Gukurahundi* upon the second major tribe, the Ndebele people in Matabeleland and Midlands provinces of Zimbabwe, between 1982 and 1987. The Ndebele constitute one in five of the Zimbabwean population. Mistrust had developed between two former war of liberation movements that were in the Zimbabwean government at independence in 1980. These are ZANU PF and the Zimbabwe African People's Union (ZAPU). The mistrust started from the time of integrating the military wings of the political parties – the Zimbabwe African National Liberation Army (ZANLA), the military wing of ZANU PF, and Zimbabwe People's Revolutionary Army (ZIPRA), the military wing of ZAPU, and the Rhodesian army – into a single national army after independence in 1980. The divisions had been created on regional grounds corresponding to the two guerrilla armies' regional patterns of recruitment and operation during the liberation struggle in the 1970s. ZAPU recruited predominantly from the Ndebele-speaking regions of Matabeleland and ZANU from the Shona-speaking regions. The divisions also rode on the back of the history of animosity between their political leaders (Alexander et al. 2000: 185–186). ZANU PF accused ZAPU of refusing to recognise the government's sovereignty and suspected it of threatening the regime security of ZANU PF. The solution lay in shooting them down and crushing their alleged leader Joshua Nkomo (Alexander et al. 2000). The 'dissidents' broke away from the national army because of what they claimed to be political bias in favour of former ZANLA cadres, especially where promotions were concerned (Tarusarira 2016). Former ZIPRA comrades in the national army were increasingly subjected to arrest, detention and harassment by the Central Intelligence Office (CIO) (Zvobgo 2009: 258). Robert Mugabe, the leader of ZANU PF, unleashed a violent attack on the Ndebele-speaking people in the Matabeleland and Midlands provinces where ZAPU dominated. Civilians were accused of harbouring dissidents. Within weeks, North Korean-trained 5th Brigade soldiers massacred and tortured thousands of civilians (The Catholic Commission for Justice and Peace 2008; Alexander et al. 2000: 181). 'Massacres, mass beatings and destruction of property occurred in the village setting in front of thousands of witnesses' (Eppel 2004: 45). To protect and save its members, ZAPU joined ZANU PF in a Unity Accord of 1987 between ZANU and ZAPU. The Unity Accord, however, managed to stop only the overt violence; it was a way for ZANU PF to ensure regime (rather than human) security. Meanwhile, regime security was concerned with the welfare, safety and protection of the ruling elite and its cronies. Its main reference points are territorial integrity, sovereignty, and state security (see Ndlovu-Gatsheni, 2003: 100–101). The report of the Chihambakwe Commission of Inquiry into the Matabeleland massacre was never made public, and the Clemency Order of 1988 – after the Unity Accord – pardoned all violations committed by all parties between 1982 and the end of 1987 (Huyse 2003: 38). That covered the Matabeleland atrocities, with neither account-ability nor apology.

Victims of violence were buried in mass graves; they did not receive decent burials (The Catholic Commission for Justice and Peace in Zimbabwe 2008: 361). The 5th Brigade, the special army created to deal with the dissident problem, insisted that there was to be no mourning. This was more stringently enforced if the victim had been killed after allegations that they were a 'sell-out'. Their relatives and friends would not be given any opportunity to mourn them. To do so would label each mourner a sell-out as well. In some instan-ces, relatives who wept were then shot. So people suffered the loss of their loved ones in silence, if not asked to take part in the killing (Eppel 2006). It was usual for victims to be buried in mass graves. At other times, people would simply be shot and left without concern for what might happen to their bodies. As a rule, the bodies were left where they had been killed, with burial forbidden, or they were buried in mass or individual graves in the villages but not in the accepted place or manner, which would entail how the corpse should lie in the grave. According to Ndebele culture, the body of the deceased is placed sideways facing the south, from where they originated. The accepted manner also entails a proper religious service and ritual, without which there would be an undignified death, severing the dead from their ancestors, their kin and their community (see Ranger 2004: 123). Where people were burned to death inside their huts the bodies were left there, while yet others were buried at 5th Brigade camps or dumped into mine-shafts (Ranger 2004: 362).

Those methods of disposal make it clear that families were left without a body over which to mourn. There was no focal point for the traditional rites, so accordingly there were no rites at all. The spirits, both of those who were buried in mass graves and of those who disappeared, remain unappeased. Yet the performance of traditional funeral rites is a pathway to healing and con-solation, whether the deaths were witnessed or whether the dead were reported 'missing or disappeared', where the 'disappeared' are people known to have been taken from their homes at night or in mysterious circumstances, or known to have been detained and never seen again (The Catholic Commission for Justice and Peace in Zimbabwe 2008: 360–1). In neither case were decent bur-ials with full traditional rites carried out, yet spirituality and communal life are at the centre of life in Southern Africa (see Hamber 2003). Spirituality plays a key role in times of sorrow and joy. Thus, in times of death and for the con-solation of the bereaved and coming to terms with the loss of a loved one, spirituality plays a pivotal role. Traditional processes of healing and reconci-liation have been advocated because they are seen as 'creating cultural social spaces that accommodate both victims and perpetrators, in facilitating the acknowledgement of wrongs and the validation of the experienced pain and loss, with a view not only to achieving the mutual coexistence of all affected parties, but also facilitating healing and consolation' (Machakanja 2010: 8).

They are consolationscapes, i.e. spaces, practices and strategies of living with loss and bereavement. Owing to the experience of *Gukurahundi*, the bereaved have been wounded at various levels. In some instances they are wounded by ignorance of what had happened to their relatives and friends; thus they have no closure, which is key to consolation. In other instances, they suffer mentally through imagining the brutality experienced by their relatives and friends. Yet no one has admitted accountability, apologised or made amends for perpetrating the violence. Thus, the Ndebele people have not become psychologically reconciled with the *Gukurahundi* experience. This chapter argues that, while space/place and time are vital for death, dying, mourning and remembrance, they remain ineffective without the ritual practices mediating the dynamics associated with death and dying. In connection with the mourning and consolation of the surviving victims of *Gukurahundi* in Zimbabwe, in the next section I consider the traditional funeral rites of the Ndebele people.

The practice of Ndebele traditional funeral rites

One way of dealing with the legacies of violence, towards healing and consolation, is to invoke the traditional idiom rooted in the idea of a moral partnership between the living and the dead (Alexander et al. 2000: 254). The relationship between the living and the dead is at the centre of African traditional spirituality. 'Religion provides support and consolation that help to overcome man's fear of the unknown, and his anxiety about the future' (Popenoe 1974: 409). The presence of a corpse during funeral rites, which is not the case with *Gukurahundi* victims, provides the survivors with the opportunity to send the dead to the world of their ancestors. Ritually, when someone has died, they lie in state at their home. Overnight, people sing and dance in honour of the deceased; the songs express the merits of the deceased and the grief of the community. The funeral wake is also an opportunity for people to talk about the individual who has died, and about death in general, in a way that consoles close family and friends.

The dead person is celebrated as the 'ever-good-one'. Hence, in Shona, they say '*wafa wanaka*' to mean that once a person dies the living stop saying negative things about them. Here, language becomes a means of consolation. The funeral is not the occasion to say negative things about the deceased. This, in itself, is consoling because it gives the feeling of assurance that the person will qualify to be an ancestor because it is the morally upright and good ones who become ancestors. For the living descendants, to be sure that the deceased will be transformed to become an ancestor is to receive the assurance of a guardian. This is comforting, healing and consoling. Because of the dreadfulness of death, people find ways to dignify death and the deceased, and sacralisation is one way to do so. Religion is thus invoked as a prime source of strength and hope. When a person dies, the Shona people might say *kuda kwaMwari* (it is God's will), or *adanwa nevadzimu* (has been called by the ancestors) (Kaguda 2012: 60). Proverbs, idioms, metaphors and euphemistic expressions play an important role in consolation. The following Shona words are cases in point: *nematambudziko* (sorry for the difficult time), *atorwa naMwari* (has been taken by God), *atorwa nevadzimu* (has been taken by the ancestors), *aiwa azorora* (has rested), *nguva yanga yakwana* (his/her time was due), *ndineurombo nekurasikirwa kwamaita* (I am sorry for the sad loss), *ndineurombo* (my condolences), *tikubatei maoko* (sorry for sad loss, accompanied by literal shaking of hands as in a greeting which signifies joining in the grief), *nedzataunganira* (sorry for the sad gathering, signifying that we are in it together), *masiiwa* (you are left behind, he/she has gone ahead of you, suggesting no need to worry because you will be reunited later on), *azorora* (has eternal rest), *awana zororo* (has found rest),*zorora murugare* (rest in peace), *nzvimbo yekuzorora* (resting place, the grave) (Kaguda 2012: 63).

These proverbial, idiomatic, metaphorical and euphemistic expressions serve to acknowledge the pain and grief of the living descendants, and to assure them that death does not mark the end of their relationship with the deceased since there is life after death. It gives explanations at the cognitive, emotional and moral level as to why the death has occurred, and in so doing that it had to be or that the person has moved to a better place or state, and that the community shares in their bereavement. It, too, has been affected. The expressions thus ultimately serve to console the family and friends. Death is presented as a peaceful and comforting sleep. It provides relief for the dying person, especially if they had been in pain with the illness that led to their death. Death is also presented as a reward or an achievement. It involves a sort of liberation in which the deceased and his/her survivors, the bereaved, will find some hope and consolation. That is why the Shona would also say *apedza pake* (has played his/her part to completion) or *arwupedza rwendo* (has completed his/her journey). Perception adjustment is thus vital to facilitate a transformation of negative feelings, pain and trauma, so the language that is deployed plays a significant part in this process.

Overnight singing and dancing before the burial are part of the funeral rites. Music is not only a means of social entertainment; it can evoke powerful emotional responses (Koelsch 2010: 131). Evoked emotions impact and enhance the subjective experience of other sensory stimuli and more specifically, music releases oxytocin which is critical to the generation of trust and affiliation (Baumgartner et al. 2006: 151; Missig et al. 2010: 2607). In the case of death, the surviving family and friends will develop the trust that all will be well despite the loss of a loved one and that the deceased is still with the survivors. Tension and pain at the overnight wake are relieved by the antics of a friend or friends (*sahwira* in Shona language, *umngane* in Ndebele language), who are honoured guests at the funeral and who had a permanent joking relationship with the deceased and their family. A *sahwira* is free to make fun of the deceased's immediate family, and so break the intensity of mourning by introducing laughter into the proceedings (Bourdillon 1976: 200). All this constitutes the process of easing the pain. It is a way of reducing the distress caused by the death and serves to facilitate consolation.

Temporally, the process of consolation starts from the moment of death. It is woven into the entire process of death and dying.

The grave-site is an important feature of dealing with the loss of a dear one with respect to healing and consolation. It provides the focal point, not only at the burial ceremony, but also for the rite of *umbuyiso* (the 'bringing back home' ceremony), through which the deceased is returned to watch over the family they have left behind. In this way, it emphasises the relationship between space/place and death (see Maddrell and Sidaway 2010). Burial is focused on the dead, but the concept of consolationscapes allows for a change in focus from the dead to the living descendants, i.e. family and friends. The complexity of the funeral rites often obscures the fact that while the focal point for the ritual is the deceased in the short run, it is for the benefit of the living in the long run. Funeral rites are intended to advance fortunes and reverse misfortunes by ensuring that the dead are properly buried to become ancestors who will watch over the living. This is not, however, to overlook the fact that the ritual serves to avoid leaving the spirit of the dead in the limbo of hopelessness, darkness and oblivion. The spirit should not be left to wander about the forest as a lost sheep, dejected and marooned, entirely out of contact with both the living and the dead (Kumbirai 1977: 125). The spirits of the people who are massacred and buried inappropriately during *Gukurahundi* are often said to be wandering about the forests and in limbo. Community efforts to conduct burials were met with resistance from the Mugabe regime, further reversing the process of consolation among the Ndebele. Through the concept of consolationscapes, we are able to focus more on the cognitive, emotional and moral meaning of *umbuyiso* in connection with consolation. In the Ndebele tradition, when someone dies it is most important that they should spend a night in their own house – specifically in the kitchen – before they are taken to be buried. Two spatial features are notable here: the grave and the home. Zimbabwean national political heroes and heroines are buried at the National Heroes Acre in Harare. However, when someone dies and is declared a national hero or heroine, they are flown to their rural home, which in most cases is outside the capital. This demonstrates the importance of home, which is often in the rural areas. Before burial, friends and family view the deceased and pay their last respects. This is perceived as an opportunity to officially say 'goodbye' to the deceased. Doing so is, in itself, perceived as consoling because it ensures closure. Closure stops the mind from continuous speculation that the 'disappeared' will return or that the deceased is not really dead and gone. The mind begins to accept that the person has truly died.

During burial, the grave is treated before or after lowering the corpse, depending on the tradition of the family. This serves two purposes, namely to protect the corpse allegedly from witches and it is a part of uniting the dead with the spirits. The family and friends then throw soil onto the coffin saying, in Ndebele, '*Hamba kuhle; usikhonzele*' (Go in peace and plead for us), as a way of bidding farewell while, in the interment service, names of all the fore-fathers are mentioned and each is asked to receive and keep the deceased, an actual committal of the soul (Bozongwana 1983). The spitting of saliva onto the soil to be thrown into the grave is another symbolic way of sending him victorious, while praise-names of the deceased spoken at this part of the service are a plea for accommodation. This ritual process is healing and brings consolation to the relatives, because the sense of having honoured the dead relative with appropriate rituals is therapeutic and cathartic. Family and friends gain the conviction and certainty that they have respected their dead, and the process is important because it ensures future protection from the dead person. It is consoling to have the feeling that the right or appropriate rituals have been performed, so that friends and family may continue to live under the sacred canopy (Berger 1967). That feeling has a healing effect. The other side of it is that if these rituals are not followed, the family and friends will live in fear that the deceased will cause misfortune amongst them.

A few weeks or months after the funeral, a ceremony known as *nyaradzo* takes place – formerly known as *mharadzamusasa* among the Shona. The meaning of the word *nyaradzo* is 'consolation', but this is not to suggest that the ceremony is the centre of consolation. It belongs to all the other ceremonies performed after a death. Its aim is to celebrate the life and to cherish the memory of the loved one. Food is prepared and served, speeches are made and poems recited, the traditional drumbeat is sounded, and even church hymns are sung. The memorial service is an occasion to sit, talk, eat, drink and bring finality to the chapter of the life of a family member or friend, who for the past weeks or months should have been sitting at the table for dinner but has not been present and will never sit there again. Everybody now knows and realises that what has happened is irreversible and all that is left are memory traces of what once was. It is a time to honour and cherish the memory of the departed, a time to celebrate their life, in a way they would have been happy with if they were alive. Nowadays, there is a mixture of Christian and African traditional rituals. Christians celebrate with sermons, prayers and Christian hymns at a memorial service. Lately, a new term for the memorial service has emerged: the 'celebration of life service'. A memorial service may be held anywhere, even at what was formerly the deceased's favourite restaurant. Funeral directors, grief counsellors, or ministers of religion may be involved in organising a memorial service, depending on either family preferences or instructions in a will (Mataranyika 2010). The memorial might be thought of as the last of the funeral rites for Christians, apart from occasional prayers for the deceased. For traditional indigenous religion practitioners, the bringing home ceremony – *kudzora munhu mumusha, kurova guva* (Shona), *umbuyiso* (Ndebele) – will follow after a year.

Thus, a year or more after the initial funeral a date is set for *umbuyiso*. Unless the dead are brought back home, the living will be unable to communicate with them. The ritual practice is meant to bring the spirit of the dead officially home and to inaugurate it as part of the ancestral guardianship over the family. While Bozongwana (1983) notes that all adults who were parents have this ceremony performed for them, Ranger (2004: 114) writes that it was not carried out for the childless, for witches or for suicides. He adds that the ritual practice also ended the period of mourning for the widow and permitted the distribution of property. Nowadays, the property can be distributed on the day of burial. This is because family and relatives live in towns and cities far away from the rural homes where funerals and burials are normally conducted, and it costs time and money to organise another family meeting for the distribution of property. Property distribution has a consoling effect. It is not uncommon to hear a woman, after receiving the jacket or shoes of a deceased aunt, saying in Shona '*Ndipo pandichatoonera vatete vangu ipapa*', meaning that it is through these items that she will 'see' and always be reminded of her aunt. Thus, the aunt is present in and through the jacket and the shoes. She is gone but present, and she is present but gone.

For *umbuyiso* the elders of the family gather together, and corn is soaked in water by the oldest member of the family while he says: 'This is your beer (referring to the deceased); we are bringing you home so that you look after the children.' On the day of the service, beer is brewed and one calabash of beer is set aside for the spirits to drink at night. At sunset, the service begins at the grave (Bozongwana 1983). The grave thus features as a central aspect of the ritual practices that are directed at consolation and healing. The centrality of the grave and its soil is demonstrated in that during pre-colonial tribal wars, the surviving warriors would bring back home some soil from the graves of those who had died in battles far afield and ceremonies would be performed around this soil. So is it when people die and are buried in towns and cities (Bourdillon 1976: 212). During the making of colonial cities, when the Ndebele people who moved away to live in cities had died, *umbuyiso* was performed in the rural areas if the person had a rural homestead. If the person had been buried in the township, his relatives would come from the rural areas, go to the grave and perform ritual acts there, talking to the grave, saying they had come to take the person's spirit to go and protect his children. They would take some soil from the grave to his homestead. Back in the rural home they would sprinkle the soil at the person's favourite resting place and perform other acts that completed the ritual (Ranger 2004: 115–116). The practice of performing ritual practices at gravesites in towns and cities remains in place in present-day Zimbabwe, considering that many people now live in cities and towns and it can be expensive to travel with the body to the rural areas for a funeral.

The importance and centrality of space and time is mediated by materiality. We have taken account of the role of the soil. It was brought back by the warriors from battle-fields, and it is taken to the rural home to be sprinkled at the deceased person's favourite resting place in the rural areas. Materiality is thus a key aspect in consolation. Whereas, in ritual theory, materials are used to facilitate attention focus during ritual practices (Marshall 2002), in addition, especially in this case, material objects have agency in themselves. It is the belief in the agency embodied in materiality such as the soil that brings about healing and consolation.

To further interrogate the role of materiality in consolation and healing, while remaining within the context of the Ndebele, let us now consider how the process of *umbuyiso* unfolds. Through *umbuyiso* a person undergoes the change of status from being one of the living community to being a family spirit (Bourdillon 1976). It is important, as in all rituals, that the correct form is followed. The descendants must do all that is required for the spirit's return and avoid everything that might offend it (Kumbirai 1977).

Some families take a goat to the grave to offer to the deceased, in order to appease them and induce them to come home. The goat is then taken back and killed for the spirits. Other families cut a branch and drag it from the grave into the home; while yet other families take beer in a calabash, sip it once as they talk and pour the libation on the grave and then walk back home singing '*Woz' ekhaya*' (come home). On the afternoon of the same day, an ox would have been slaughtered, skinned and put in the hut for the night. The goat is killed, roasted and eaten the same night by the family, who pause only to sniff tobacco at regular intervals. This is a communion service in which all members of the family can speak to the ancestors and request whatever they want from them. The rest of the meat, a calabash of beer and snuff are left in the hut for the spirits to feast. Various materials are notable here: the grave, animals such as goat and ox, beer, calabash, and snuff. As materials with agency, they give the survivors a sense of certainty about their beliefs regarding everything surrounding the dead. From a psychological perspective, these materials might be perceived as transitional objects, defined as a designation for any material to which an infant attributes a special value and by means of which the child is able to make the necessary shift from the earliest oral relationship with mother to genuine object-relationships (Winnicott 1953). In this case the materials at play serve to take care of the gap created by the death of a loved one. Thus, some objects must take the place of the deceased and bring structure, safety and assurance that life has not been necessarily ruptured and one still has protection. The feeling that one has protection facilitates consolation and healing.

In the morning, beer is spilled on the ground as a libation while people dance and sing: '*ubaba makeze ekhaya*' (father should come home). An ox is roasted and eaten, and beer is served as well. Some of the beer offering is left overnight for the relatives after all the other people have gone. One member of the family is then chosen (usually a boy child) to represent the dead man, particularly if the deceased was a medium or the boy was named after the deceased. The immediate implications in the communion sacrifice mean the readmission of the recently deceased into the home in another form and capacity, and it underlines the importance of the belief in immortality and the appeasement of spirits. The sharing of the sacrificial meal secures permanent acceptance for a belief in the communion of the living and the dead. The purpose of the sacrifice is to express or establish a relation of harmony and unbroken fellowship leading to looking after the family. With the presence of this child, the people can safely say '*baba vari pamba*', meaning that father is here at the homestead. To facilitate communication, the spirit is given an animal as its host. It is then installed on the day of *umbuyiso*. Whenever there is illness in the home, the head of the kraal will go to the ox very early in the morning and will speak vehemently to the spirits as he kneels beside the animal (Bozongwana 1983). That is why the Shona also call the ceremony '*kutamba n'ombe*' (to dance to the sacrificial beast) (Kumbirai 1977). In strong language, the spirits are told to stop molesting and causing illness to the family: 'If you don't protect your children, what is your work then?', the man would say to the spirit. Consolation can also thus be understood as the ability to deal with the gap created by the deceased. It can be understood as the ability to re-establish or restore the structure disrupted by the death of a dear one. The entire ritual process can be well-located with both the functional as well as the transformative dimension of ritual practices. Functionally, the former structure of the family and community is restored once *umbuyiso* has been performed. There is, thus, a movement from structure (the one that has been disrupted) to anti-structure (caused by death) and back to structure, after the ritual practice, in this case *umbuyiso* (see Turner 1974; 1976; 1982). However, the restoration of structure is facilitated through a transformative process at two levels. First, the new structure is not qualitatively the same as the previous one, because the deceased is said to be present but not in the same way as before, and the relationship with them is now different. Second, the surviving family and friends have been transformed cognitively, emotionally and morally as a result of the experience of the death and ritual practices. Thus, the functionalist and transformative dimension of rituals exist simultaneously or are co-present in the ritual practice of *umbuyiso*. Moreover, both the functional and transformative elements in the death ritual facilitate consolation.

Following *umbuyiso*, an annual drink, food and snuff offering is made even if there is no illness. We have noted that *umbuyiso* takes place after a year, and this is a stipulated time that is central to the ritual form, for the sake of its operational efficacy. Frequent holding of the ritual slowly heals the wounds of family and friends. It is, however, my observation that with time the frequency of ritual practice decreases. A successful year characterised by the fertility of land, livestock and also human beings is attributed to the co-operation and direction of the ancestral spirits and, as such, all that man has is ordered by their governance. This brings about the feeling that the surviving relatives and friends are not alone. It is a confirmation that the dead are present and active in people's lives as they had been during their own lifetime. Having outlined the role of traditional funeral rituals, it is in order now to examine how *Gukurahundi* has violated and desecrated the traditional funeral rites of the Ndebele and how this has stifled the healing and consolation among the survivors.

Gukurahundi and the desecration of religio-cultural values

As has been explained above, the ancestors play a special role in the world of the living. Consolation comes about as a result of knowing that the ancestors are watching over the family. In their physical absence, they become spiritually present through ritual practices, not gone but simply transformed. Bereavement occurs when the dead are cut off from family and friends. To gain consolation, survivors try to counter this absence through bringing back the spirit of the dead. Thus, the search for consolation is a counter-process to death. For an ancestral spirit to play its role effectively, and thus to facilitate consolation, it needs an honourable funeral followed by the traditional *umbuyiso* ceremony. A spirit that has not been honoured through these rites becomes an angry and restless spirit, in a scenario that can bring bad luck to the family and the community at large. Bad luck can be exhibited in children's bad behaviour, children's failures, illness, droughts, floods and crop failure, among many other things. It is not surprising that the droughts of 1991–1992 and 1994–1995 provided the impetus for a critique of moral and religious appraisal in Matabeleland. The subsequent remedies were religious, particularly the desire for cleansing the metaphysical and physical traces of violence, during the liberation and post-independence wars (Alexander et al. 2000: 264–265). When a problem is attributed to an angry spirit, it is normally interpreted as follows: 'The spirit of so and so is in the wilderness, she or he is wandering and was not brought home to watch over his or her family' (Eppel 2006: 264–266). Carrying out the aforementioned rituals has the simultaneous effect of healing and consoling the individuals and the community. It heals both the wandering spirit's immediate family and also the whole community, since the ritual has value insofar as it is part of the community value system.

Consolation facilitates the general health and well-being of an individual, but the individual is connected to the community; their health and well-being cannot be maintained alone or in a vacuum. Healing and consolation rituals and community are vitally linked (Somé 1999: 22). Rituals are the most ancient way of binding a community together in a close relationship, thus co-presence is an important aspect not only of the ritual practices but also of the consolation process. No wonder then that, among the Shona people, the consolation of friends and relatives is referred to as *kunyaradza vafirwa* (to console those who have lost a family member or friend). This implies that the process of healing and consolation after a death requires community assistance inasmuch as it is also for the individual. The pain of the grieving family and friends is a community pain. This togetherness becomes a support system for the directly affected to bear the pain of losing someone close to them. This resonates with the concept of *ubuntu* in African philosophy, which states that a person is a person because of other people, translated as *umuntu ngumuntu ngabantu* in Ndebele and *munhu munhu navanhu* in Shona. In this context, *ubuntu* is a medium of consolation. Rituals have been the most practical and efficient ways to stimulate the consolation and safe healing required by both individuals and the community. Ritual has been the way of life of the spiritual person because it is a tool to maintain the delicate balance between the body and the soul, which gets disrupted by traumatic experiences such as death. It is so deeply connected to human nature that whenever it is missing, as was the case in the *Gukurahundi* era, when relatives and friends were unable to perform the funeral rituals, there will be a lack of consolation, transformation and healing. Rituals therefore should not be seen as empty, old-fashioned, irrelevant, boring, dark and pagan, with no place in modern times, but as transforming, essential, healing and consoling (Somé 1999: 141–146).

To deny traditional burial is to subvert cultural values, which both offends and disturbs the survivors. Not performing these rituals, as was the case with *Gukurahundi*, makes survivors feel vulnerable to danger within the fortune-misfortune complex (De Craemer et al. 1976: 463). As in many parts of Africa, and as exemplified by the Ndebele, people cannot live without a sacred canopy (Berger 1967), known as *mumvuri* (shade) in Shona, under which there is a hope of deliverance and protection from evil in all its different forms (Anderson 2001: 233). Berger (1967: 26–27) proposed another category opposed to the sacred, in addition to the profane, namely chaos. The sacred cosmos, characterised by order and harmony emerges out of chaos. The sacred cosmos transcends and includes man in its ordering of reality, thus providing man's ultimate shield against terror of anomie. To be in good standing with the sacred cosmos is to be protected from chaos while to be otherwise is to be abandoned on the edge of the abyss of meaninglessness. Interesting enough is the observation that the English word 'religion' derives from the Latin, meaning 'to be careful'. Berger further writes that the sacred is potentially dangerous although its potency can be domesticated and harnessed. The religious man is therefore careful about the dangerous power inherent in the manifestations of the sacred where behind it lies an even greater danger, which is losing all connection with the sacred and being swallowed up by chaos, hence all the nomic constructions designed to keep terror at bay. The sacred narrative guides and informs the life of many African people, not excluding the Ndebele. Their world is a 'continuum between the visible and invisible worlds, and mankind shares its environment with spirits of some type who influence mundane transactions and with which direct communication is possible' (Ellis and ter Haar 1998: 179). To remain in communion with the dead is thus to avoid chaos and danger. When safety, order and good fortune are ensured, consolation follows.

Gukurahundi desecrated the religio-cultural systems and needs surrounding the dead, not only by refusing proper burial, but also by forbidding mourning, and forcing people to take part in disrespectful behaviour such as dancing and singing on the shallow graves of the newlymurdered, and leaving bodies in pain of death. Cultural systems serve to meet the human needs for meaning. They provide meaning of life at three levels: cognitive, emotional and moral. At the cognitive level they explain why things are as they are. It is not enough to know 'what is'; people also need to know what they are meant to feel about this picture of life and universe presented to them, whether to be awestruck, amazed or fearful, or to be hopeful, joyous or welcoming. Cultural systems supply the emotional meaning, and offer guidance on what to feel and under what circumstances. They provide moral meaning by helping people understand why things are as they are when judged not simply from a disinterested scientific standpoint but from an interested and partial perspective of human desires, hopes and expectations. A successful cultural system, therefore, is one 'that tells people what it is that they should think, how they should feel, on what basis they should judge others as well as themselves, together with what actions they ought to perform to attain salvation, peace or enlightenment' (Campbell 2007: 167). The desecration of cultural systems through *Gukurahundi* ruptured the framework of meaning upon which the survivors hang in their daily life, in a world characterised by fortune and misfortune. Following death, the traditional funeral rituals described in this chapter are directed towards restoration of the cognitive, emotional and moral meaning.

Conclusion

In the context of conflict and violence, dealing and coming to terms with the tragic past is a search for consolation and healing. Traditional funeral rites and rituals are a key part of consolation. While in other contexts, especially Western contexts, physical and psychological woundedness and death are dealt with by psychologists and psychiatrists, in many African contexts there are rituals at the centre of consolation. The case of *Gukurahundi* demonstrates the vitality of those rituals. Religious rituals, under which funeral rites fall, give a framework within which to explain problematic things that happen in people's lives and sometimes provide practical responses to that. Rituals help in understanding and coping, and have the psychological function of integration and equilibrium. In instances following death, coping is consolation. Through their ritual practices, people are able to accept situations such as death. Religion and rituals guide the socio-economic and political aspects of life, hence religion and spirituality are not abstract principles but something practical. They are profoundly integrated into social and technical life. To think of them as disappearing, as was argued by advocates of secularisation thesis, including the Indian historian and diplomat K. M. Pannikar who argued that religion would decline in Africa and Asia following the end of Western colonial rule (see Meyer 2004: 452), is superficial and without depth.

References

Alexander, J., McGregor, J., and Ranger, T. O. 2000. *Violence and Memory – One hundred years in the 'dark forests' of Matabeleland*. Oxford and Portsmouth, NH: James Currey and Heinemann.

Anderson, A. H. 2001. *African Reformation: African Initiated Christianity in the Twentieth Century*. Asmara: Africa World Press.

Baumgartner, T., Lutz, K., Schmidt, C. F., and L. Jancke. 2006. The emotional power of music: how music enhances the feeling of affective pictures. *Brain Research* 1075, 151–164.

Berger, P. L. 1967. *The Sacred Canopy: Elements of a Sociological Theory of Religion*. New York: Open Road Integrated Media.

Bourdillon, M. 1976. *The Shona Peoples*. Gweru: Mambo Press.

Bozongwana, W. 1983. *Ndebele Religion and Customs*. Gweru: Mambo Press.

Campbell, Collin. 2007. *The Easternization of the West: A Thematic Account of Cultural Change in the Modern Era*. Boulder, CO: Paradigm.

De Craemer, W., Vansina, J. and Fox, R. 1976. Religious movements in Central Africa: a theoretical study. *Comparative Studies in Society and History* 18/4, 458–475.

Ellis, Stephen and ter Haar, G. 1998. Religion and Politics in Sub-Saharan Africa. *Journal of Modern African Studies* 36/2: 175–201.

Ellis, S., and ter Haar, G. 2004. *Worlds of Power: Religious Thought and Political Practice in Africa*. New York: Oxford University Press.

Eppel, S. 2004. 'Gukurahundi': The Need for Truth and Reparation, in *Zimbabwe: Injustice and Political Reconciliation*, ed. Brian Raftopoulos and Tyrone Savage. Cape Town: Institute for Justice and Reconciliation, 43–62.

Eppel, S. 2006. Healing the dead: exhumation and reburial as truth-telling and peacebuilding activities in rural Zimbabwe, in *Telling the Truths: Truth Telling and Peace Building in Post-Conflict Societies*, edited by T. A. Borer. Notre Dame, IN: University of Notre Dame Press.

Hamber, B. 2003. Healing, in *Reconciliation after Conflict: A Handbook*, edited by D. Bloomfield, T. Barnes and L. Huyse. Stockholm: International Institute for Democracy and Electoral Assistance, 77–89.

Hayner, P. B. 2011. *Unspeakable Truths: Transitional Justice and the Challenge of Truth Commissions*. London: Routledge.

Huyse, L. 2003. Zimbabwe: why reconciliation failed, in *Reconciliation after Violent Conflict: A Handbook*, edited by D. Bloomfield, T. Barnes and L. Huyse. Stockholm: International Institute for Democracy and Electoral Assistance, 34–39.

Kaguda, D. 2012. Death and dying: An analysis of the language used in coping with death in the Shona society. *Journal for Studies in Humanities and Social Sciences* 1/2, 57–68

Koelsch, S. 2010. Towards a neural basis of music-evoked emotions. *Trends in Cognitive Science* 14, 131–137.

Kumbirai, J. 1977. Kurova Guva and Christianity, in *Christianity South of the Zambezi*, edited by M. F. C. Bourdillon. Gweru: Mambo Press, 123–130.

Machakanja, P. 2010. *National Healing and Reconciliation in Zimbabwe: Challenges and Opportunities*. Wynberg: Institute for Justice and Reconciliation.

Maddrell, A. and J. D. Sidaway. 2010. Introduction: bringing a spatial lens to death, dying, mourning and remembrance, in *Deathscapes: Spaces for Death, Dying, Mourning and Remembrance*, edited by A. Maddrell and J. D. Sidaway. Farnham: Ashgate.

Marshall, D. 2002. Behavior, belonging and belief: a theory of ritual practice. *Sociological Theory* 20/3, 360–380.

Mataranyika, P. 2010. Memorial: last of funeral rights, http://www.financialgazette.co.zw/memorial-last-of-funeral-rites/, April 23, 2010, accessed 21. 04. 2018.

Meyer, B. 2004. Christianity in Africa: from independent to pentecostal-charismatic churches. *Annual Review of Anthropology* 33, 447–474.

Missig, G., Ayers, W. L., Schulkin, J and Rosen, J. B. 2010. Oxytocin reduces background anxiety in a fear-potentiated startle paradigm. *Neuropsychopharmacology* 20, 858–865.

Ndlovu-Gatsheni, S. J. 2008. Who ruled by the spear? Rethinking the form of governance in the Ndebele State. *African Studies Quarterly* 10/2&3, 71–94.

Ndlovu-Gatsheni, S. J. 2003. Dynamics of the Zimbabwean crises in the 21st century. *AJCR*, 3/1, 99–134.

Philpott, D. 2006. *The Politics of Past Evil: Religion, Reconciliation, and the Dilemmas of Transitional Justice.* Notre Dame, IN: University of Notre Dame Press.

Popenoe, D. 1974. *Sociology.* New York: Prentice Hall.

Ranger, T. O. 2004. Dignifying death: the politics of burial in Bulawayo. *Journal of Religion in Africa* 34/1–2, 110–144.

Somé, M. P. 1999. *The Healing Wisdom of Africa: Finding Life Purpose through Nature, Ritual, and Community.* New York: Tarcher/Putnam.

The Catholic Commission for Justice and Peace in Zimbabwe. 2008. *Gukurahundi in Zimbabwe: A Report on the Disturbances in Matabeleland and the Midlands, 1980–1988.* New York: Columbia University Press.

Tarusarira, J. 2016. Revisiting Zimbabwe's celebratory history in pursuit of peace and reconciliation, in *Sources and Methods for African History and Culture: Essays in Honour of Adam Jones,* edited by K. Werthmann, G. Castryk and S. Strickrod. Leipzig: Leipziger Universitätsverlag, 613–628.

Turner, H. 1969. The place of independent religious movements in the modernisation of Africa. *Journal of Religion in Africa* 2/1, 43–63.

Turner, H. 1974. Liminal to liminoid in play, flow, and ritual: an essay in comparative symbology. *Rice University Studies* 60/3, 53–92.

Turner, H. 1976. Ritual, tribal and catholic. *Worship* 50, 504–526.

Turner, H. 1982. Religious celebrations, in *Celebration: Studies in Festivity and Ritual,* edited by V. W. Turner. Washington, DC: Smithsonian Institution Press, 201–219.

Winnicott, D. W. 1953. Transitional objects and transitional phenomena. *International Journal of Psychoanalysis* 34, 89–97.

Zvobgo, J. M. C. 2009. *A History of Zimbabwe, 1890–2000 and a Postscript, Zimbabwe, 2001–2008.* Newcastle-upon-Tyne: Cambridge Scholars.

9 Love the dead, fear the dead

Creating consolationscapes in post-war northern Uganda

Sophie Seebach

Introduction

Let us take a walk through Lily's compound.[1] Lily is in her sixties and lives a short motor-cycle ride from a trading centre, which was once a massive refugee camp. We are in Gulu District in Acholi, northern Uganda, and it is dry and sunny. On the neatly swept compound stand three thatched mud huts, new thatch waiting in upright bunches, for Lily's young relatives to arrive and start working. A gentle breeze plays with the red dust, but it is not enough to dry the sweat pearling on our foreheads, as Lily slowly leads us around the compound and points out the graves of her dead.

Sources of pain as well as of consolation, the graves of her parents, her daughter, her daughter's infant twins, and other relatives are clearly visible. Her daughter, mother, and others are buried under simple mounds of soil and scattered bricks. Her father lies in an elaborate cement grave with his name etched into it. And the twins are buried, befitting their status as *lotino jok*, or spirit children, in pots in the middle of the compound. Here Lily lives with the dead, and they help her through grief and hard times.

In Acholi, the dead belong in the home, where they are buried next to houses and huts, placed according to age, gender, and manner of death. There are multiple funeral rites, sometimes spanning years, which ensures a continued ritual interaction with the dead. Thus, the day-to-day interaction between the living and the dead does not cease at the point of death; indeed, it continues as they proceed to share the same space.

Surrounding Lily's compound is the wider landscape, the bush, which in northern Uganda has become forever associated with its history of insurgency and guerrilla warfare. Unsettled and angry spirits 'roam the bush' and attack unwary people fetching water or walking along paths. The bones of dead soldiers and rebels are still turning up from time to time, their spirits hungry for revenge, but seemingly uncaring whether or not their victims had anything to do with their death. Therefore, while there might be consolation to be found in the landscape in Acholi, there is also fear.

1   All informants have been anonymised to protect their identity.

Acholi consolationscapes

This chapter is about the 'consolationscapes' of Acholi.[2] It explains how the structure of the landscape, in which people find themselves, aids their consolation in the wake both of personal loss and of the larger societal devastation of civil war. The dead and the land, the soil, are intimately connected in Acholi (Meinert and Whyte 2013). Placing the dead in a certain way in the soil of the ancestors is a way for the living to create order, and through this order to create the possibility of consolation. This chapter explores how consolation is found through tying the dead to the land and to the home. Yet, it is not only the bones and the memories of the dead which are embedded in the landscape; it is their very being, their spirits, and their agency. Acholi death rites mould the landscape; the landscape is indeed shaped into 'consolationscapes', or, more specifically, into 'kinship consolationscapes'.

These landscapes invite a process of consolation, not just on a personal level, but also on a communal level, following a harrowing civil war. The gathering of the living to commemorate and honour the dead not only sends a powerful signal to the dead that they remain a part of the family; it also serves to gather the extended family network and reaffirm connections among the living – connections which are vital for consolation.

Yet the actions taken towards this end are not simply in order to create a space for the living and the dead to reside in peaceful unity. They are also, as we shall see, an attempt to control something fundamentally uncontrollable: to tame the dead, who are to be feared as much as they are to be loved. Living with the dead poses a constant danger, since living in such close proximity to the dead signifies an obligation, which, if it is not upheld, might have catastrophic consequences.

It might seem surprising that these spirits play such a large role in daily life, considering that Acholi is predominantly Catholic. Yet many Acholis do not see an issue in attending Church and adhering to the word of God, while conducting rituals to honour or soothe spirits, and burying loved ones next to the home. Indeed, there seems to be a particular kind of Acholi Christianity, which is an amalgam of Christian and 'traditional' beliefs – an 'Acholicized Christianity', to paraphrase Behrend (1999b: 22). Herein, the traditions from an era prior to the advent of Christianity remain vital to the ongoing welfare of the dead as well as of the living. Some groups, such as Pentecostal Christians, are less accepting of this blending of tradition and ritual, and were often, among those who were interviewed, the people who claimed to know the least about Acholi burial rites. And, though they still bury their dead in the home, they often claim not to know why this is the norm, merely stating that it is 'Acholi tradition'. Therefore, even in such families, the dead remain a physical as well as a spiritual presence in everyday life.

---

2   This chapter builds upon twelve months of fieldwork, conducted in Acholi between 2011 and 2014.

In the consolationscapes of Acholi, people work with their grief, while at the same time, they maintain ongoing, active, relationships with the beloved and feared entities for whom they are grieving. Klass has stated that grieving is inter-subjective (2014: 6), and in Acholi this grief, and the consolation with which it is met, takes place within a collective of the living and the dead. And this transcendent relationship is deeply embedded within the Acholi landscape.

The LRA war and the displaced dead

One of the reasons for the massive presence of the dead in Acholi is the area's recent history of civil war. The following paragraphs will not delve deeply into the complicated political history of Acholi; instead, they will merely outline a history of what has come to be known as the LRA war, and how it affected dealings with the dead.

Twenty years of war

When Yoweri Museveni, the current president of Uganda, came to power in 1986, tensions already brewing between north and south were exacerbated. Museveni, a southerner, ousted the remnants of the former regime's troops from Kampala; they fled to their home region of northern Uganda and formed the basis of a rebellion. Several rebel armies were assembled (Allen and Vlassenroot 2010; Behrend 1999a), but none was more successful than Joseph Kony's 'Lord's Resistance Army' (LRA). Kony, an Acholi and self-proclaimed *ajwaka* (a spirit medium or witchdoctor), initially claimed to want to protect the Acholi people from Museveni's regime. However, he quickly turned his rebel forces on his own population, abducting thousands of children to swell his ranks of child soldiers, and brutally punishing any civilians suspected of collaborating with the government (Atkinson 2010, Dolan 2009, Eichstaedt 2013, Finnström 2008). The Ugandan army too were guilty of atrocities, killing civilians, looting cattle, destroying villages, and at times deliberately leaving the civil population to the mercy of the rebels (Dolan 2009: 44).

From the mid-1990s onwards, up to 90 per cent of the population – approximately two million people – were relocated to camps for internally displaced persons (IDP camps; see Dolan 2009: 107, Tapscott 2015: 3). There people lived and died under often horrible conditions, the victims of illness, starvation, poor hygiene, and rebel attacks. Those who died were buried in the camps, as their relatives were prevented from leaving the camps to bury them in their ancestral homes. Thus, their graves were crammed in between huts on strangers' land, sometimes without the proper rituals, in direct contravention of Acholi tradition and custom (Meinert et al. 2017).

When the LRA definitively pulled out of Uganda and peace was established in Acholi in 2006, the families of the dead were left with the massive task of not only returning home themselves to rebuild their homes and lives, but also to bring their dead with them. According to Acholi custom, the dead are to be buried in their ancestral home, as their spirits prefer to linger among their living relatives (Meinert and Whyte 2013: 175). People's places in social life are very much structured around age and gender, and a person's place in this social structure is not negated by their death. The dead are buried in the home, securing their continued existence among their living relatives, and their place in the family is reflected in their place and manner of burial. The graves of women are usually placed on the side of the cooking hut, which contains the stove. Odoki, an Acholi Catholic Bishop with a doctoral degree in theology, explains that this is due to the woman's role as the cook and caregiver of the family: 'The hearth-stones and fire are the symbol of livelihood at home. The woman is therefore buried near the fire-place in the house where she can spiritually continue to play her role of motherhood' (1997: 38). Men are placed at the opposite side of the house. The younger a person is, the closer to the walls of the hut they are buried. Children, and the spirits of children, are in need of much love and attention, and so they need to be close to their living relatives. Infants feel this need the most and thus they are buried within the wall of the hut. Furthermore, the bereaved perform between two and three burial rites, spanning a period of months or years, depending on the age of the deceased and the manner of their death.[3] These rites serve to satisfy the spirit, and show the family's continued devotion to the deceased, to gather the extended family for communal mourning, and eventually to tie the dead to the home.

This intricate system was disrupted brutally by the war, but the end of hostilities allowed the bereaved to attempt to remedy the situation by reburying the dead. Before we explore the consolationscapes of Acholi, and why reburials are significant to their creation, let us briefly explore the phenomenon of reburial in Acholi.

Going home: Post-war reburials

The reburial of a loved one is both a joyous and a tense occasion. On the one hand, the family are bringing a deceased relative home where they belong, but on the other, they are disturbing spirits and manipulating remains, which is always a perilous venture.

Spirits are a significant part of Acholi life, and keeping them happy is vital. There are many different kinds of spirits, but here it will suffice to mention two; the *tipu* and the *cen* (Meinert and Whyte 2013: 175). *Tipu* is the spirit which dwells in us all and which, upon our death, is released from our bodies to roam around, eventually to settle in its ancestral home with its living relatives. The *cen* is quite different. *Cen* is a vicious creature; it is what you become if, for example, you have died a violent death or if your grave has

---

3   See Odoki (1997) and Seebach (2016) for in-depth descriptions of Acholi burial customs.

been desecrated (Finnström 2008: 159). And in the process of reburial, you must always be careful not to turn the *tipu* of the deceased into a *cen*.

A reburial is sometimes instigated by a wish on the family's part, because of problems with unsettled spirits or because of a desire to be near the dead. But often, the event that forces the family to act is when the land-owners on whose land the dead are buried want the graves removed, perhaps because they want to develop the land, sell it, or farm it (Seebach 2016: 169). I took part in four reburials, three where a single person was reburied, and one reburial of two adults and two infants. Depending on the distance, and the resources of the family, the remains are moved either in a coffin or wrapped in a blanket. Once, I attended a reburial where the remains were transported in a tattered white bag (see Meinert and Whyte 2013). Normally, remains are then carried to their new resting place either by hand, on the back of a bicycle or a motorbike, or in a car or a truck. At the old grave, a goat is slaughtered as part of an effort to calm the spirit down and convince it that what is happening is good and not a malicious attempt to tamper with the grave. Once home, the family buries the dead and celebrates the return of their loved one.[4]

A reburial is expensive, which is one of the reasons why many families have yet to rebury their deceased relatives. It is a venture carried out mainly by family and friends, and thus priests are rarely involved. Indeed, the world of spirits seem much more at the fore in reburials than the Christian beliefs to which most people in Acholi adhere. In their deeds and words, people show their love for the spirits, but also the apprehension and fear, which characterises relationships with the dead in Acholi.

Reburying Opio and Ocen

This ambiguous relationship with the spirits is well exemplified in the story of Lily, whom we met at the beginning of this chapter.[5] Sitting in the cool shade of her hut, she tells us about her life. Her hair is cropped close to her head, and while her left eye meets mine directly, her right eye is half-closed, its iris milky-white and unseeing. She tells me how her daughter Santa died in the IDP camp during the war, along with her two twin babies, Opio and Ocen.[6] The reburial of the twins, she says, was much more complicated than that of her daughter:

---

4   For a more detailed account of a reburial, see Meinert and Whyte (2013) and Seebach (2016).
5   The initial contact with Lily was made in collaboration with Tove Nyholm, as research for the exhibition *The Lives of the Dead* at Moesgaard Museum, Aarhus, Denmark.
6   Twin boys in Acholi are always named Opio and Ocen (and if they are girls, Apio and Acen). Opio is the first to be born, Ocen, whose name hints at a special connection to the spirits, is the second.

[Twins] are special. They disturb people. If you play around and don't respect them, there will be sadness and death. [Opio and Ocen] were buried in the camp where people could not see the graves. Our neighbour had placed some bricks on top of the graves, and when we had to rebury them, we couldn't find them. So we considered only taking Santa. But we went to the *ajwaka* to ask her advice, and she recalled where they were, and the owner of the bricks had to remove all the bricks [...] Then we could remove the bodies. We brought a sheep to kill when we removed the twins. We dug down and removed the pots [in which the twins were buried] and [killed] a sheep next to the hole [...] First we brought the children home, then the mother; the twins were respected even more than the mother.

When dealing with twins, the procedure of reburial is slightly different from other reburials.[7] Twins are buried in pots, instead of coffins or blankets, and, instead of the usual goat, a sheep is killed. In this example you see Lily's desire to do right by her dead, but also the apprehension felt towards them, and the need to rebury them lest they bring 'sadness and death'.

So, why focus on reburials when exploring the consolationscapes of Acholi? Because reburials are one of the most potent ways in which families create an environment that can allow them to heal. The burial rites conducted after a death are also a way to create consolationscapes; when families perform the multiple funeral rites, and when they place the dead in specific places in the home compound, in accordance with their place in the social structure of the family, a kind of order is restored. By positioning the dead into their place in the social structure, some of the chaos of death is becalmed, and the living are able to grieve in peace in the knowledge that the relationship between the living and the dead is as it should be.

Communal consolation

The consolationscapes created through reburials happen in response to a crisis much greater than that of an individual death. Beyond the grief and need for consolation that the death of a loved one entails, reburials are an answer to the collective trauma that is civil war. By physically moving the dead from the wrong place to the right one, the bereaved shape the space around them into the home it should be, and would have been had it not been for the war. During an interview with an old woman, whose family had re-buried several relatives, I asked if it made a difference to bring the dead home

---

7 Twins, as well as children born with what is considered physical 'defects' (from small cosmetic irregularities to severe physical handicaps), are considered *lotino jok*, 'spirit children' or 'children of the gods'. *Lotino jok* are treated with added respect, and must also be buried (and reburied) slightly different to 'normal' people, as their spirits are more volatile and powerful.

from the camps. She answered: 'It is important. It shows that all the people are one people. People should be buried at home. This person [who was recently reburied] is part of our family. We can't separate. If you made a new grave somewhere, you would have to bury more people so it is not alone.' It is evident in this and other interviews that this notion of a family as being 'one people' is not just an emotional one; it is significant that they are also united in space.

Here we have, to quote Sahlins, a 'mutuality of being'; a kinship system where people 'are members of one another, [and] participate intrinsically in each other's existence' (2013: 2). The individual is not in focus; instead, it is the desire to preserve the community which takes precedence; it is in the community (of living and dead) that safety and consolation are to be found.

The correct placement of bodies in space makes the spirits, as well as the living, content, and it orders social life. It illustrates what is important in life: family and ancestral relationships, as well as the fundamental connection to the ancestral land. Indeed, this last point is significant. As Meinert and Whyte put it, '[t]he reburial of the bones in the ancestral land creates material continuity between land and people, as well as between the living and the dead, which makes new life possible, and at the same time confirms a family's claim on the land' (2013: 175). There is material gain and comfort to be had in securing the rights to one's ancestral land. This significant connection between land and family, ancestors and descendants, is important not just in Acholi but in other Luo communities as well.[8] Parker Shipton describes this relationship among the Kenyan Luo:

> Burial of the dead – male and female – within homesteads provides crucial fixed points on the landscape for the reckoning of personal, familial, and political identities and allegiances. Luo people make it clear that they look upon graves and the old homestead sites of their forebears as their anchors – in time, in space, and in culture and society.
>
> (Shipton 2009: 14).

Indeed, in Acholi, the significance of ancestral land and of family ties is not to be underestimated. Knowing that these things remain secure and preserved for the future is a powerful component in grief work. The communality of the reburial is also underlined by the fact that the ritual itself is a communal effort. Through the ritual, the family gathers in one space, allowing members to share the burden of the reburial, financially as well as emotionally.

In Acholi, the relative safety of the home compound often stands in contrast to the surrounding landscape. This landscape is, perhaps, the opposite of

---

8   The Luo peoples, of which the Acholi are one, are a large Nilotic ethnic group, stretching over northern Uganda, eastern Congo, south Sudan, Ethiopia, Kenya, and Tanzania.

a 'consolationscape'. And it stands as a threat to the ongoing grief work, of which reburials and the recreation of the home is part.

The threat of the bush

In rural Acholi, you are always surrounded by the bush. In the wet season it is lush and green, and in the dry season it turns golden and crisp. Here there are snakes, wild bees that make delicious honey, and giant edible rats, which are chased out with fire and killed with bows or spears. And it is also the domain of the rebels. Though the LRA have now left Uganda, you still say, of those young people who were abducted and have never returned, that they 'are in the bush'. Indeed, *alum*, the Acholi word for bush or tall grass is also the word used for the rebels, and if children are born in the bush, they are sometimes named Alum, in accordance with the Acholi practice of naming children after the circumstances of their birth.

This unwieldy landscape provides a stark contrast to the consolationscape of a well-tended compound. The dead of the bush are also very different from the dead who rest, mostly peacefully, in their family homes. The spirits of the murdered, the suicides, and those, whose graves have been disturbed, roam the bush. So too, do the spirits of those who were killed in the bush and whose remains have never been brought home and buried. If these spirits come across a living person in the bush, they might attack them and cause madness or even kill them. Therefore, you should take precautions if you think you might be in danger. As one woman told me: 'If you see bones or a body in the bush, you should put a leaf on them and say "I bury you" [...] If you do not, the *cen* might think you killed it and attack you.' While this symbolic burial will not leave the spirit at peace, it should divert its wrath away from you, allowing you to escape.

Hallam and Hockey have described how sites that hold an association with death and the dead can have a powerful influence on the living. Sites like these, can 'through fleeting or permanent association with the dead, [...] evoke profound emotion by acting as potent reminders of particular persons and the condition of human mortality' (2001: 83). While the associations created by the compound and the bush respectively are quite different, strong family bonds spanning the chasm between life and death versus uncontrolled death and uncontrollable spirits, it is nonetheless not difficult to understand why even the somewhat controlled environment of the home compound needs to be tamed.

Managing the uncanny

One reason for the fragile nature of peace in the home compound lies, I argue, in the fact that the bush, as a concept and a fact, always threatens to invade. The bush is, to borrow a word from Sigmund Freud (2003), *uncanny*. It is at once familiar and unknown; you have to traverse it on a daily basis, yet on a daily basis it threatens to turn on you. The vengeful *cen* might be brought home by an unwitting traveller or with a scarred and traumatised returnee from the LRA. Memories of coming home after the war to over-grown and ruinous compounds, which have, quite literally, been reclaimed by the bush, still haunt peoples' memories, as do the memories of the war in the form of land wrangles with neighbours. During the encampment, the bush blurred the line between the home and the surrounding wilderness, and between neighbouring compounds, creating the opportunities to encroach on neighbours, causing strife and conflict.

The bush thus comes to stand as a constant uncanny potentiality; an ever-present danger, which threatens to invade the lives of the living. And though the dead are most welcome in the home, indeed they are invited in to be loved and taken care of, there is no escaping the fact that they will always be associated with the bush, and with the uncontrollable forces which dominate it. Freud, inspired by Schelling, defines the uncanny, or the *unheimlich* as 'everything that was intended to remain secret, hidden away, and has come out into the open' (Freud 2003: 132). And while the dead in Acholi were never meant to be hidden away, they are preferred in a certain place, and a certain state of unthreatening benevolence. And their inherent associations with something ultimately untameable render them inherently uncanny. Therefore, the relationship between the living and the dead, with whom they share their daily life, is characterised by a constant effort to keep the dead content. By keeping the dead happy, they come to terms with their place in the family and as a result the living, too, can achieve con-solation to some extent.

Creating consolationscapes through reburial

While the Acholi landscape might, both deliberately and through coin-cidences, be littered with 'deathscapes' (Maddrell and Sidaway 2010), the consolationscapes of Acholi are carefully and purposefully created. Verdery (1999) described in vivid detail how the moving and reburying of dead bodies had reconfigured space and time during the dismantling of former Yugosla-via. Moving the dead, in this case, became a way to claim land, to draw up boundaries, and to send powerful political messages to opponents. These strategies can provide a lens through which to view the post-war reburials happening in Acholi; and can indeed remind us that while the relocation of the dead in Acholi is in large part about creating consolationscapes, it is also about small-scale politics, power, and the right to land. Nevertheless, in a way, these issues become central to the very act of consolation.

Creating the future

When actively reforming the landscape, for example through reburial, people create landscapes that facilitate consolation. Furthermore, reburial creates a space in which the community can move forward and create a common future. As Meinert and Whyte state:

> Reburying relatives on family land is a kind of timework that encompasses past, present and future by materialising and localising the kinship relations that make a home. By interring the remains in the soil that is claimed as ancestral, people make continuity with the past; they enhance a feeling of security in the present by diminishing worries about things left undone and dissatisfied shades; and they anticipate a future of fruitful family unity in and on the earth of a place.
>
> (2013: 189)

Drawing on Flaherty's concept of timework (see Flaherty 2011), Meinert and Whyte show how the reburial of loved ones reshapes not only the past into the shape it should have had, had it not been for the war; it also creates the foundation for a different and better future. The central act of home-making provides great comfort. When we visit people who returned from the camps in recent years, it quickly becomes clear how gratifying it is to reclaim the home. 'We live according to relationships', said one old man, who had reburied several dead relatives. To him, it was vital that people be buried where they have family ties, because these relationships last far beyond death. Thus, the kinship consolationscape that is the Acholi home is literally built upon the dead. The dead become one with the soil. Beans dry in the sun on the flat surfaces of the cement graves, and dead children are buried in the walls of the homes, to receive, and provide comfort.

As such, the creation of consolationscapes comes to be about much more than individual grief and consolation; it is about families and communities jointly creating a space, which allows for a coming to terms with the past, in order to plan for the future. Communal spaces of grief are not unique to Acholi (see for example Francis et al. 2001, Hallam and Hockey 2001, Petersson 2010, Tylor 2002). Yet the hands-on way they are created, and the fact that they are such an integral part of the home itself, lends a specific flavour to the Acholi consolationscapes, which sets them very much apart from the cemeteries we are used to in Europe and the US. 'Here, the dead are not dead', as one Acholi man told me. The dead are not someone you visit in their resting place on birthdays and holidays. They are always there, sharing everyday life with the living.

The comfort and discomfort of continuing bonds

Klass writes that 'the most common source of solace comes in the continuing bonds the living maintain with the dead' (2014: 11). In Acholi, the matter of continuing bonds transcends generations. As the dead are woven into the landscape of the ancestral home, they are woven into their rightful place in the family history, and in the future of the family. Through this, the living are also able to see their place in this family structure, now and in the future, when they too will die and become ancestors. A general comment on the nature of consolation might as well have been written about the Acholi: 'in sorrow many people find consolation in the sense that they participate in something that transcends present space and time' (Klass 2014: 9).

For Lily, these bonds manifest themselves not only through the physical presence of her relatives' graves but also through her meetings with them in her dreams. Sometimes, her dead daughter Santa comes and sits by her bed, keeping her company through the night. Sometimes Santa runs away from her; those dreams make Lily sad. The twins are more active participants in their grandmother's life:

> They are watching the family. When I feel ill, I'll go and ask for help, and they make me healthy. After two or three days, I will find that I am healthy. They heal me and protect the home. They hear my voice. If you left them, and didn't rebury them, they would bring sickness. You have to respect them. I don't ask Santa, only the twins.

Lily's relationship with the dead shows that these bonds are just as complicated and multi-faceted as the bonds between the living. They are characterised, among other things, by love, devotion, duty, trepidation and fear. They are, in other words, the kind of continuing bonds between individuals that make up our social lives. There is great consolation to be found in the notion that the dead remain with us. While the body may perish, the spirit lives on and continues to play a part in the family's social network. Just like the bonds that we share with our living relatives, such relations are complicated, multi-faceted, and sometimes problematic.

As Lily's story made clear, the relationship between the living and the dead is not unequivocally a positive one. The dead, even those you protect and whose graves you maintain, are potentially dangerous. And though a landscape in which the dead and the living exist side by side might bring consolation to some, it can be hard for others, who might not appreciate the constant reminders of deceased loved ones. And the task of reburying, during which you have to handle the remains of the dead with your own hands, can be heart-wrenching.

The dark sides of Acholi consolationscapes?

> It is a sad occasion to see your father or brother being uprooted. It is painful to bring back these memories that you're trying to forget. We still have so many relatives lying in the camp, and I am afraid of what might happen [if they are not reburied]. But I am sure that we will be asked to remove them soon, but it worries me that we have no money for the goats to do the reburials properly [...] I have created my own cemetery a little distance from the house. We used

to bury people close to the home, but it pains me to see the graves every day and be reminded of those I have lost.

So an old man called Boniface told me of his relationship to his dead relatives. While many might derive comfort from the close proximity to the dead, and the unity of the living and the dead this entails, Boniface was saddened by their nearness. Moreover, he was fearful of not being able to live up to his responsibilities to the dead. One who famously argued for the existence of a 'dark side' of clinging on to the dead was Sigmund Freud. In 'Mourning and Melancholia', Freud argues that for the mourner to obtain closure and establish a healthy relationship with what they have lost, they have to sever the relationship with the deceased (1917/1957: 244). As Freud writes: 'Reality-testing has shown that the loved object no longer exists, and it proceeds to demand that all libido shall be withdrawn from its attachments to that object' (ibid.). Willerslev rephrases this far from self-explanatory prose by suggesting that the bereaved must '[kill] off the memories of attachment to the lost love object as a means of re-establishing mental health and returning to life' (2013: 86), in order to avoid the state of 'melancholia' (i.e. depression).

I am far from agreeing with Freud, and in general, as Francis et al. (2001) and Klass (2014) demonstrate, the world seems to have moved beyond this way of considering mourning. Nevertheless, what Boniface demonstrates above is that some people might nonetheless feel trapped in the continuing bonds between the living and the dead. The obligations that such bonds entail can be both a source of consolation, in that the relationship to the deceased remains beyond death, and also a burden; especially when you might not feel able to live up to them.

Since the dead are so intertwined in the landscape of the home, as they are in Acholi, these complicated relationships are always present. There is no escaping them when they are part of the home itself. One woman, who had lost many people dear to her, among them her two sons, had this to say when I asked her how it felt to live in such close proximity to the dead:

> I am not happy. I am still young, and not supposed to have death so close. Sometimes, when I sit alone, I can still cry when I look at [the graves]. There are too many dead people... my children, my husband... Because of this, I have resolved to plant trees to hide the graves from sight.

In this case, she is in fact planning to mould the landscape in order to make consolation possible. But contrary to the usual way of doing this in Acholi – keeping the dead close – she wants to shield herself from them.

In a comment on Sahlin's *What Kinship Is – And Is Not*, which I briefly mentioned above, Janet Carsten commends him on his descriptions of the 'mutuality of being' and his descriptions of the importance of kinship. But she also notes that kinship relations also have in them the potential to turn sour, for relations to turn upon one another in jealousy and anger (2013: 247). Indeed, without drawing directly upon his notions, Carsten's conclusions share DNA with Freud's descriptions of the uncanny; the *heimlich* turned *unheimlich*, the familiar, the safe and the known, suddenly turned threatening and unfamiliar, precisely because of its nearness. The proximity of the dead in Acholi is what brings comfort, to the living as well as to the dead, but it also underlines their inherently threatening nature.

Conclusion

Once, as I sat with a group of older Acholi women, they asked me about Danish burial rites. I told them how hospital staff and undertakers handle the body, how it is most often cremated, and how we bury our dead in cemeteries, sometimes far from home. 'Ah, your custom is better', one woman said with a nod. When I asked why, she answered:

> Because when they have taken the body from you, you don't feel the pain [...] I feel it is better, because you just get the body from the hospital, you take it for prayers, and then you burn it. So it is better because it does not give you that more grief and anger as when you see it at home. There is a lot of pain associated with seeing the body.

In her mind, the Danes' ability to burn their dead and place them far away from the home meant that we must not feel grief. Grief itself is tied to the physical presence of the body.

In Acholi, in fact, grief without a body is not uncommon; such is the case with the many families who are still missing members who have been taken by the LRA. However, perhaps what one might call a kind of 'productive grief', which is a precursor to consolation, *does* need the presence of a body and a physical resting place. This is why reburial, in Acholi, is so vital for the consolation process. When the living and the dead are present in the same physical space, the common kinship consolationscape, only then can they properly console each other.

Yet, when the home compound is infused with the remains of the dead and the landscape is moulded around the relationship between the living and the dead, the dead might also pose a threat to the well-being of their living relations. The ongoing relations with the dead becomes a matter of keeping the dead close, but not *too* close, of controlling them, in order to live with them. By placing them correctly in the landscape, the living relations hope to create a situation that allows for both individual and communal consolation, all the while placating the spirits of their loved ones. Love and fear thus mingle in the continued creation of consolationscapes in Acholi.

References

Allen, T. and Vlassenroot, K. 2010. Introduction, in *The Lord's Resistance Army. Myth and Reality*, edited by T. Allen and K. Vlassenroot. London: Zed Books, 1–21.

Atkinson, R. 2010. *The Roots of Ethnicity: Origins of the Acholi of Uganda*. Kampala: Fountain Publishers.

Behrend, H. 1999a. *Alice Lakwena and the Holy Spirits: War in Northern Uganda, 1985–97*. Oxford: James Curry.

Behrend, H. 1999b. Power to heal, power to kill: spirit possessions and war in Northern Uganda (1986–1994), in *Spirit Possession: Modernity and Power in Africa*, edited by H. Behrend and U. Luig. Oxford: James Curry, 20–33.

Carsten, J. 2013. What kinship does – and how. *HAU: Journal of Ethnographic Theory* 3(2): 245–251.

Dolan, C. 2009. *Social Torture: The Case of Northern Uganda, 1986–2006*. New York: Berghahn Books.

Eichstaedt, P. 2013. *First Kill Your Family: Child Soldiers of Uganda and the Lord's Resistance Army*. Chicago, IL: Lawrence Hill Books.

Finnström, S. 2008. *Living with Bad Surroundings: War, History and Everyday Moments in Northern Uganda*. Durham, NC: Duke University Press.

Flaherty, M. G. 2011. *The Textures of Time: Agency and Temporal Experience*. Philadelphia, PA: Temple University Press.

Francis, D., Kellaher, L., and Neophytou, G. 2001. The cemetery: the evidence of continuing bonds, in *Grief, Mourning and Death Ritual*, edited by J. Hockey, J. Katz, and N. Small. Berkshire: Open University Press, 226–246.

Freud, S. 1917/1957. Mourning and Melancholia, in *The Standard Edition of the Complete Psychological Works of Sigmund Freud, Volume XIV (1914–1916): On the History of the Psycho-Analytic Movement, Papers on Metapsychology and Other Works*. London: Hogarth Press and the Institute of Psycho-Analysis, 237–258.

Freud, S. 2003. *The Uncanny*. Translated by D. McLintock. London: Penguin Books.

Hallam, E. and Hockey, J. 2001. *Death, Memory & Material Culture*. Oxford: Berg.

Klass, D. 2014. Grief, consolation, and religions: a conceptual framework. *OMEGA*, 69(1), 1–18.

Maddrell, A. and Sidaway, J. D. 2010. Introduction: bringing a spatial lens to death, dying, mourning and remembrance, in *Deathscapes: Spaces for Death, Dying, Mourning and Remembrance*, edited by A. Maddrell and J. D. Sidaway. Farnham: Ashgate, 1–16.

Meinert, L. and Whyte, S. 2013. Creating the new times: reburials after war in Northern Uganda, in *Taming Time, Timing Death: Social Technologies and Ritual*, edited by D. R. Christensen and R. Willerslev. Farnham: Ashgate, 175–193.

Meinert, L, Willerslev, R., and Seebach, S. 2017. Cement, graves, and pillars in land disputes in Northern Uganda. *African Studies Review* 60(3): 37–57.

Odoki, S. O. 1997. *Death Rituals among the Lwos of Uganda: Their Significance for the Theology of Death*. Gulu: Gulu Catholic Press.

Petersson, A. 2010. The production of a memorial place: materialising expressions of grief, in *Deathscapes: Spaces for Death, Dying, Mourning and Remembrance*, edited by A. Maddrell and J. D. Sidaway. Farnham: Ashgate, 141–159.

Sahlins, M. 2013. *What Kinship Is – And Is Not*. Chicago, IL: University of Chicago Press.

Seebach, S. 2016. *The Dead Are Not Dead: Intimate Governance of Transitions in Acholi*. Unpublished PhD Thesis, Aarhus University.

Shipton, P. 2009. *Mortgaging the Ancestors: Ideologies of Attachment in Africa*. New Haven, CT: Yale University Press.

Tapscott, R. 2015. The government has long hands: community security groups and arbitrary governance in Uganda's Acholiland. *The Justice and Security Research Programme Paper* 24.

Tylor, T. 2002. *The Buried Soul: How Humans Invented Death*. Boston, MA: Beacon Press.

Verdery, K. 1999. *The Political Lives of Dead Bodies: Reburial and Postsocialist Change*. New York: Columbia University Press.

Willerslev, R. 2013. Rebirth and the death drive: rethinking Freud's 'Mourning and Melancholia' through a Siberian time perspective, in *Taming Time, Timing Death: Social Technologies and Ritual*, edited by D. R. Christensen and R. Willerslev. Farnham: Ashgate, 79–98.

10 'It's God's will'

Consolation and religious meaning-making after a family death in urban Senegal

Ruth Evans, Sophie Bowlby, Jane Ribbens McCarthy, Joséphine Wouango and Fatou Kébé

Introduction

This chapter explores consolation and religious meaning-making after a family death in the Muslim-majority context of urban Senegal. Little empirical work has been conducted on geographies of loss in Muslim-majority contexts and published in English to date. Yet research in Muslim-majority societies can destabilise normative and homogenising understandings of Islam and of Muslims (Mills and Gökariksel 2014; Falah and Nagel 2005). Indeed, Wikan's (1988) comparative research in two Muslim-majority communities in Egypt and Bali highlights the cultural diversity in the expression of religious practices, which shapes and organises responses to loss. Recent geographical work on religion has explored the intimate space of the body and the ways bodily performances, for example in prayer, ritual, dress and so on, are shaped by different sets of formal or informal rules, norms and expectations (Gökariksel 2009). However, such work has not explored issues of consolation in Muslim-majority societies.

As Maddrell (2009) has observed, while bereavement, like religion, is often compartmentalised in designated spaces and times, it is nevertheless ongoing and pervades everyday life. Yet few studies to date have explored the everyday significance of death within the familial and social context. Research in the Minority world[1] has been heavily driven by a focus on individualised psychological processes of what may be understood to be 'normal grief' and any indications of need for professional interventions. In studying death in Senegalese families, we have drawn instead on Klass' (1999) notion of 'responses to death' in order to move away from the dominant Minority world discourse of the experience of the death as an individualised journey of 'grieving' (Rosenblatt and Bowman 2013). Cacciatore and De Frain (2015) suggest that,

1   We use the terms Majority and Minority worlds to refer to the 'Global South' and 'Global North' respectively, to acknowledge that the 'majority' of the world's population and land mass are located in the former and to highlight the extent to which contemporary global power and knowledge is located in contexts that are historically and numerically specific. We recognise, however, that this dichotomy is problematic and risks homogenising and obscuring extensive diversities.

in African societies generally, it is the community consequences of death that are centralised.

Recent literature from the Minority world has called for more attention to be paid to the material dimensions of death and absence (Hockey et al. 2010; Maddrell and Sidaway 2010) and pointed to the significance of continuing bonds with the deceased (Klass et al. 1996; Steffen and Coyle 2011; Silverman, 2017). Such continuing relationships with those who have died may be expressed through activities, thoughts and practices and are shaped by relationships in life, the nature of the death and socio-cultural differences (Klass et al. 1996; Howarth 2007). Klass (2014) suggests that the most common source of solace comes in the continuing bonds the living maintain with the dead.

This chapter discusses the findings of our qualitative research on responses to death, care and family relations in urban Senegal, where the vast majority (94 per cent) of the population identify as Muslim. A minority of the population are Christians (4 per cent), mainly Roman Catholic, or animists or other religions (2 per cent) (ANSD 2013). Sufi-Islam is often practised through affiliation to a variety of hereditary brotherhoods (four main ones) which have a significant political presence, while Catholicism is part of the French colonial legacy. Islamic and Christian practices regarding burial, funeral, mourning and inheritance have mingled with indigenous cultural practices that vary according to ethnic group (Sow 2003).

Our research[2] aimed to investigate the material and emotional significance of a death of a close adult relative for family members of different genders and generations. We identified a purposive sample of 30 families living in Dakar and Kaolack who had experienced an adult relative's death in the previous five years. In-depth interviews were conducted with 59 family members including 30 children and youth (aged 12–30) and with 23 key informants, in addition to four focus groups. Participatory workshops were held with 45 family participants and with 29 government and non-governmental representatives and Muslim religious and local leaders to gain feedback on our preliminary findings and the policy implications.

Drawing on Klass' (2014) understanding of consolation and religious solace, this chapter explores participants' narratives of the death of a relative and the role of co-presence and religious practices in enabling meaning-making[3] and the (sometimes constrained) expression of grief and continuing bonds, which provided consolation to bereaved family members. We first discuss the conceptual framing and how this relates to responses to death.

---

2   See http://blogs.reading.ac.uk/deathinthefamilyinsenegal and Evans et al. (2016) for more information.

3   We understand 'meaning-making' here in terms of hermeneutic approaches which put human meaning-making centre stage, as (often taken-for-granted) efforts towards sense-making. See Ribbens McCarthy et al. (in press) for further discussion.

Conceptual framing: Consolation and religious solace

Klass (2014: 3) suggests that three elements are helpful when seeking to understand the religious worlds of bereaved people:

> First, encounter or merger with transcendent reality; that is, the sense that there is something beyond our mundane existence that we can, at least for moments, experience as an inner reality. Second, a worldview that provides the cultural narratives in which our individual life narratives are nested. A worldview gives meaning to the events and relationships in our lives. Third, a community in which transcendent reality, our worldview and our own experience is validated.

He suggests that this triune structure is evident in many religious traditions, including Islam. For example, 'Allah is the God who can be found but who cannot be understood by human intelligence; the Prophet Mohammed was given the revelation to which humans should conform their lives; and the Ummah is the community of all those who submit to Allah' (ibid.: 3). As we demonstrate, our findings support Klass' argument that consolation happens within each of these elements.

Klass (2014) understands consolation as social and inherently inter-subjective, in the same way as grief. Solace is characterised by 'the sense of soothing' and being comforted in the midst of sorrow's hopelessness and despair: 'Solace alleviates, but does not remove distress' (ibid.: 6). Tracing the etymology of consolation-related words, such as *comfort* from the Latin *fortis*, meaning 'strong' or 'powerful' and the prefix *com* from *cum*, meaning 'with', highlights the relational nature of the verb to comfort, that is, to strengthen or find strength together (ibid.: 6). Swedish and German interpretations (*Trostworte*) reveal that comfort words, consoling words, are those which restore trust: 'Solace is found then, within the sense of trusting or being connected to a reality that is outside the self' (ibid.: 7).

Klass (2014) shows how this understanding of solace can be found in human relationships, cultural resources, religious meanings and continuing bonds. For example, solace is found through human relationships which are characterised by non-judgemental co-presence, in which the 'helper' is open to the pain, becomes available and is present (ibid.: 8). This aspect relates most clearly to the second element in the understanding of religion outlined above. Solace may also be found in cultural resources; 'in sorrow, many people find consolation in the sense that they participate in something that transcends present space and time' (ibid.: 9). This relates to the first and third elements outlined above. Further, religious faith and rituals often provide consolation through a sense of religious connection to a transcendent reality beyond our mundane existence. This may include prayers to the saints for the wellbeing of the dead found in some Christian and Muslim traditions, practices such as lighting memorial candles, holding church services (Klass 2014), saying aloud the *kaddish* prayer in the Jewish community and the annual marking of the *yahrzeit* anniversary of the death (Silverman, 2017), and in our research, holding a recital of the Koran and making offerings to others in memory of the deceased. Thus, as Klass suggests (2014), the religious sense within solace is multi-faceted.

Following this conceptual framing, in the next sections, we explore our empirical findings from urban Senegal. We focus on participants' narratives of a family death, the spiritual meaning-making they engaged in and the role of religious practices and co-presence in providing consolation.

Narratives of the death and religious meaning-making

Interviewees had diverse kin and in-law relationships with the deceased family member. The largest proportions of the sample had lost a husband, a mother or a father. Most of these relatives had died in middle or older age, although several had died in their thirties. Interviewees often talked at some length and in some detail about the events surrounding their relatives' deaths. However, participants rarely gave a detailed medical narrative or explanation of the 'cause of death', in the way that might be common in the Minority world (Valentine, 2008) (see Table 10.1). Many young people did not know the cause of their relative's death and both adults and children mainly talked of symptoms rather than specifying diseases. Only one interviewee explicitly said that they had a '*certificat de décès*', an official death certificate for the deceased issued by the town hall and few mentioned the '*constat de décès*', or death notice, which is mandatory and usually issued by a doctor.

The most frequently mentioned biomedical references concerned chronic illnesses such as high blood pressure/hypertension, diabetes and cancer. Interviewees rarely mentioned discussing their relative's diagnosis or cause of death with doctors or healthcare professionals providing treatment. As Boubacar, an NGO worker, commented about his brother's illness: 'those are questions we don't ask'. A minority of participants made reference to 'mystical illnesses' and/or reported that they had taken their relative to a traditional healer, although the use of traditional healers may have been under-reported, given the intimate and private nature of patient–healer relationships and therapeutic landscapes (Bignante, 2015).

The majority of interviewees gave detailed narratives of the events leading up to their relative's death, the moment of death and how they heard the news. This suggests meaning-making around the death that may not be intended as a 'causal' explanation in more Anglophone terms (Sogolo 2003). Several interviewees whose relatives had died in old age thought that their relative had felt that they were going to die or wanted to die, which was often manifested in a reluctance to seek medical treatment. Some interviewees referred to older relatives' sense of premonition and preparation for death. Samba (aged 51), whose grandmother had died in her 90s, drew on a notion of 'death sickness':

Table 10.1 Five narratives of relative's illness and/or death

"Explanation' of relative's death	*Number of interviewees*[*]
Medical term used e.g. diabetes, hypertension, cancer	22 (37%)
Description of physical symptoms	17 (29%)
Did not specify/did not know	16 (27%)
Religious explanation	15 (25%)
Accident or assault	4 (1%)

[*] Due to our methodology, researchers did not ask all participants about medical explanations for their relative's illness. Findings are based on relatives' self-reports and so more stigmatised illnesses or causes of death, such as HIV, mental illness or suicide may be under-reported. Two family members' accounts are provided of one relative's illness/death, although their responses were not necessarily consistent. Religious explanations have been counted independently of the other categories, as these were sometimes drawn on in addition to other accounts, at later stages of the interview.

> Maybe she had 'death sickness'. We told her on that day that we were going to take her to the hospital. She used to say 'I'm satisfied because I've seen my grandchildren.' (...) That's why she refused when we wanted to take her to the hospital. (...) She knew she was going to die.

Interviewees' accounts of unexpected deaths of younger relatives who were only ill for a short period of time or who had been killed violently or in an accident, were experienced, understandably, as a great shock. Such deaths were more difficult for family members to share with close relatives and to accept and make sense of, in comparison to the deaths of relatives who had chronic illnesses or who died in old age. In the case of a family whose relative had been killed in a violent assault, news of the death was met with fear. Some interviewees appeared to make sense of such unexpected deaths by drawing on events leading up to the death that they felt provided a warning or sign of the potential danger facing their relative. This suggests that notions of destiny or fate, inflected with religious and cultural meanings, may be drawn on for meaning-making when faced with sudden, untimely deaths. Prothmann's (2017) research with young men in urban Senegal, for example, suggests that fate was often seen as 'God's will, ineluctable and transcendent' and explained with the Wolof phrase, *'Ndogalu Yàlla'*, translated here as the 'Judgement of God', whereby life is regarded as predestined and caused by God, who had already chosen an individual's day of death.

Many interviewees drew attention to deaths which had occurred during religious festivals, such as Tabaski, Tamkharit, Magal, Gamou, Ramadan (Muslim festivals) or Easter (Christian festival), or on Fridays (significant to Muslims). As Abdoulaye's (aged 30) account suggests, the religious significance of the day his mother died and his religious faith helped him to accept her death:

It was hard but I trusted in God and what's more, she died on the day of Tabaski. Everybody wants to have this day [to die] so I thank God. She did a lot of good things and I know only God can reward her. (...) She died at prayer time. (...) I thank God; it's Divine will. (...) Everyone has their day.

The use of the phrase 'trust in God' here illustrates Klass' (2014) idea that religious solace is found by trusting in a transcendent reality. The phrase, 'Everyone has their day' highlights the sense of theistic predetermination of the day of death (Prothmann 2017).

Some interviewees' narratives of the death included religious practices such as reciting the Koran and placing it on a relative's head at the moment of death. Investing the moment of death with religious significance appeared to provide solace and help participants accept the death. Supporting Klass' (2014) understanding of the three elements of the religious worlds of bereaved people, an encounter with a transcendent reality at the moment of death, the religious worldview that gives meaning to the death, and the faith community, all seemed to validate people's experiences.

In contrast, some younger women expressed fear when their relative died at home. Diami (aged 26) was sweeping the house at the time and went to check on her mother. She felt 'so afraid' that she did not think her mother had died, but when she realised, felt alone at this moment: 'I felt very bad because I only had my mother.' As others have noted (Walter 1999; Dunn et al. 2015), death can be seen as a threat to social cohesion, as well as breaking and disrupting familial relationships and support networks, with the risk of isolation. Jacquemin (2010) suggests that rather than being a choice, solitude and isolation in Africa (more than elsewhere) appears to be a sign of a loss of social status and support, and implies greater suffering (see Evans et al. 2017 for further discussion).

A striking idiom used frequently by participants was 'it's God's will'. When asked about their feelings about a relative's death in interviews, a quarter of participants (see Table 10.1), mainly middle and older generation adults, made reference to *God's will* in making sense of the death or said that it was God's decision that their relative should depart the earth at that time. Abdoulaye (aged 30, Muslim) was one of the few young people who drew on this religious framing of the death, when asked, 'Can you tell us what you felt at the time of your mother's death?' 'It's like everybody. It's very hard but I left everything in God's hands. (...). It's God that brought her onto earth and God who took her back, and nobody will escape that day.'

The frequent use of such religious refrains, including, 'it's God's will' [Wolof: *ndogalou yalla;* French: *c'est la volonté de Dieu*], 'it's Divine Will' [Wolof: *ndogalou yalla la*; French: *la volonté Divine*] and 'it's in God's hands' [Wolof: *si lokho yala*; French: *c'est entre les mains de Dieu*] suggests that religious beliefs and worldviews associated with Islam and Catholicism, inflected with Senegalese cultural framings, provided the predominant frames of reference that enabled participants to engage in meaning-making about the death. Such socio-cultural framings appeared to help participants accept the death and resign themselves to its inevitability, providing consolation. For example, Cheikh, an older man (aged 77) whose mother had died, had sleeping problems, an embodied manifestation of his grief, but appeared to find solace and peace in his religious beliefs: 'It was hard. I couldn't sleep at first but in the end I realised it was God's will'. Similarly, Safietou, a mother (aged 50), made sense of her son's death (when he was 23) and her sense of powerlessness in the face of this through her religious faith:

> Being Muslim, I can only trust in God. It's He who'd given him to me, it's He who took him. I can do nothing. There's a moment when you hold back; you pray for his soul and appeal to Him (…). Ultimately you pull yourself together.

Many family members felt powerless when faced with the inevitability of death, but appeared to be comforted by prayer and the wider religious and cultural context that gave meaning to their experiences. N'diouga (63-year-old widower) explained that the word *'Mounieul'* in Wolof, meaning endurance, was often used in relation to the need to accept death:

> Like they say in Wolof, *'Mounieul'* [you must endure/ persevere]; that is, you must be aware that everything perishes so it's not worth creating a drama. You must remain strong; everyone does, yes, even women.

De Klerk (2013) observes a similar notion in Tanzania that refers to a coun-selling/advice-giving practice that is particularly apparent at funerals and is used to comfort bereaved relatives; *oyegumisilize* means 'forgetting' or 'enduring', from the Kiswahili for 'being tough' and 'being healthy'.

Alongside this strong cultural imperative to 'persevere' and 'remain strong', a Roman Catholic priest suggested the Wolof value of *'Natu/Nattu'*, meaning a test of faith, or in some interpretations, a curse from God (Prothmann 2017), was important in understanding a person's ability to accept a death in the Senegalese cultural context. Thus, being able to accept the death through a religious framing could be regarded as a judgement of a person's depth of faith and commitment to their faith community and religious practice:

> You feel the depth of the person's faith. It's when we're tested. It's for that, in Wolof they say *'Natu'*. *Natu* is something we say that measures your faith. God does it to measure your faith; to see how far your faith goes; the depth of your faith.

For many participants, religious beliefs, practices and belonging to a religious community were intrinsically bound up with a relational sense of self, familial and communal responsibilities and the need to survive and 'keep going'. This supports Klass' (2014) relational understanding of religious solace, characterised by the inter-weaving of religious faith, practice and community.

It was also apparent, however, that some people felt they could not accept the death, even though this might be going against the teachings of religion. As Ibrahima (aged 44) commented about his mother's death two years previously:

> (...) this gap we're still feeling until now. Sorrow; I'm even ashamed to think of her to tell you the truth, because I still haven't accepted this death. I pretend that she's still here. That's what helps us to keep going.

Unusually among our interviewees, 'keeping going' here appears to be linked to pretending the death had not happened and seeking comfort in the sense of a continued presence of the deceased as a living person, which perhaps a religious perspective would discourage.

Co-presence and 'getting by'

Family and community solidarity was central to helping bereaved relatives to 'keep going' and 'get by' after a family death (Ribbens McCarthy et al., forthcoming). To understand its particular salience in the Senegalese context, it is vital to recognise the economic precarity of life for most people and the crucial nature of family and household ties to everyday economic survival in a context in which there is a minimal welfare state (Evans et al. 2016; see also Randall and Coast 2015). Cultural values of *Solidarité* [in Wolof: *Dimbalanté*] and *Teranga* – terms that refer to mutual support and norms of unquestioning hospitality to any visitor – are socially approved and considered a widespread feature of life in Senegal (Bass and Sow 2006; Gasparetti 2011), although we were also told that these norms were weakening in the city. The significance of these values and relationships was highlighted by Boubacar (aged 44): 'Without the family, we're nothing. Without friends, we're nothing. Without neighbours, we're nothing.'

Consolation may be found simply in the demonstration that people of significance to the bereaved offer their condolences and share their loss. This sharing may be expressed through silent co-presence at their home during the mourning period. As N'della (aged 19), whose father had died, explained: 'The family is very important to me... It's because if I cry and I see them beside me, it's as if I have everything I need beside me.' Jackson (2004) argues silent co-presence in response to suffering may be especially valued in African societies as a form of healing that helps to restore the social world. Silent co-presence may also have been more important historically in the UK amongst those with few resources (Strange 2005).

Co-presence may also be expressed through face-to-face talking and phone calls. Salimata, a 62-year-old widow whose son had died in the month prior to the interview, said:

Members of my family can't afford much but they all love me. The day he died they were all there; even those from the villages far away came. Those that couldn't come, they phoned me. Even today, there were two that phoned me.

Such (sometimes silent) actual or virtual co-presence may be understood as expressing care for the bereaved person but also as affirming that 'family' support is and will continue to be available. The bereaved person is recognised as part of the reciprocal web of care that can be offered by family relationships now and in the future, providing a sense not only of sharing the loss, but also of existential security. Solace is thus found in the widespread cultural narrative of *solidarité* which gives meaning to the death and strengthens reciprocal kinship and community support networks.

A sense of consolation and shared grief was particularly evident through the support provided by neighbours and friends during the funeral, which was sometimes spread over multiple days. For Muslims, the burial must take place within a day of the death and funeral gatherings often follow the burial, usually at the home of the deceased.[4] Relatives living some distance away, thus, may not always be present in the immediate aftermath. Neighbours, associates and friends living nearby as well as household members are available to help with the practical funeral arrangements, supply much needed material support in the form of food or money, in addition to providing emotional and spiritual support and advice:

> When a person dies all the neighbours, all your neighbours, your friends, everyone comes to share your pain. Everybody comes and gives something to help you with the costs, we really live as a family, even neighbours are part of the family.
>
> (Djibril, aged 42, after the death of his aunt)

As de Klerk (2013) observes in Tanzania, comforting at the funeral may involve listening to the bereaved family members' story of the deceased's last days and narrating stories about others who have experienced loss and survived, so that the bereaved do not feel alone in their loss. Offering condolences and adhering to reciprocal norms of *solidarité* may provide solace by reaffirming the cultural narratives through which family members make sense of the death and relationships in their lives, thereby strengthening social cohesion (Klass 2014; Dunn et al. 2015).

Tensions were also apparent, however, in some participants' narratives about religious and familial injunctions not to cry 'too much' or for 'too long', underpinned by understandings of the death as 'God's will' and the need to 'stay strong' and 'keep going':

---

4   See Evans et al. (2016) for more extensive discussion of funeral procedures.

You shouldn't exaggerate because everything has a limit...religion doesn't tolerate a person crying for so long [during the funeral period]... Of course, religion allows us to cry but if you persist, it's like calling into question Divine will.

(Head of district, Guédiawaye)

This need to limit crying was articulated by both adults and young people, sometimes linked explicitly to religious beliefs, while at other times, it was linked to practical demands to continue everyday routines or not to upset children or other family members. Wikan (1988) suggests that suffering and sadness may be regarded as contagious and detrimental to all; to the self, other people, as well as to the soul of the dead, which leads to strong social sanctions and the need to look after the living. As de Klerk (2013: S489) observes, 'enduring' and being 'gently counselled that one's situation is entirely normal' silences emotions and creates social cohesion.

Indeed, many interviewees emphasised the importance of 'keeping going' and 'getting by' [French: *se débrouiller*] to ensure personal, household and family survival, which we suggest is linked both to the religious worldview and to families' material circumstances within poor urban neighbourhoods. In this respect, we noticed among many interviewees a hesitancy to acknowledge that changes had occurred following the death. When probed further about changes in their lives, many nevertheless pointed to many significant, often negative changes. When we discussed this hesitancy in the feedback workshops, the explanations offered by participants concerned the need to maintain the family's material wellbeing and status in the same way as before, alongside religious understandings. Thus Toufil, a young widow (aged 25) who had moved back to live with her natal family following her husband's death, said: 'It's badly regarded if someone dies and one says that there have been changes. So you will try to do what that person there did for you.' Acknowledging changes appeared to reflect poorly on the ability of surviving relatives to 'persevere' and continue to support the family in the way their deceased relative had done.

Other participants alluded to religious understandings. A young woman (aged 20) said, 'When someone dies, one shouldn't say negative things about them. It's a sin.' Another young woman (aged 21) in a different workshop linked the reluctance to talk about changes to the belief that the death was 'God's will'. Thus, continuing to grieve and acknowledging that the family's circumstances had become more difficult following the death could be regarded as a failure of the 'test of faith' [Wolof: *Natu*] and an inability to provide for their family.

The difficulty of acknowledging publicly that a death had led to emotional distress and created economic or practical strains in the ability to support bereaved family members may reinforce the pressure to 'be strong' and limit a bereaved person's ability to express their grief. But, as we have indicated, the co-presence and support of family and community members and interwoven religious beliefs and meaning-making about the death offered crucial sources of consolation. We acknowledge, however, the specific nature of our sample, recruited through local intermediaries, and the inevitable cultural repertoires of presenting a 'good death' that takes place within the community, near relatives (Ndiaye 2009), both of which are likely to have influenced participants' narratives. As we discuss in the next section, consolation was also often found within religious practices of remembrance and the expression of continuing bonds. In these ways, grief and difficulties could be acknowledged, albeit within a wider religious and cultural context that sought to regulate emotions and foster social cohesion.

Continuing bonds and practices of remembrance

Continuing bonds, that is, 'the relationship that individuals, communities and cultures maintain with those who have died' (Klass 2014: 11), are inherently relational. Perhaps because their expression affords a degree of agency to both the living and the dead, continuing bonds and practices of remembrance may offer considerable consolation to bereaved family members. The act of remembering can be seen as a way of thinking, style of behaviour or conduct which affects people's feelings, actions and physical wellbeing (Klaits 2005; de Klerk 2013). In our research,[5] prayers were by far the most significant practices of remembrance and continuing bonds for the dead. As Saer, a young Muslim man (aged 22) said, 'If somebody's died, you can only say prayers for them'. For Muslims, daily prayers included praying for all deceased relatives:

> I wake up every morning at 5am. I do my ablutions and I pray for him and my deceased relatives, and for all the other deceased Muslims.
>
> (Nogaye, widow, aged 46)

Thus, the practice of praying connected the living to the dead, and the particular deceased relative often became identified with the wider set of deceased family members, and with the dead more broadly. An imam explained that in Islam, prayers are regarded as for the benefit of the dead:

> It's said in religion that when someone dies there are angels who come to ask him questions and he'll answer (...). These prayers we do for him can allow him to answer easily.

Djibril (aged 42) suggested that the practice of daily prayers encouraged reciprocal continuing bonds and mutual care between the living and the dead:

---

5   It is beyond the scope of this chapter to discuss widowhood mourning practices. See Evans et al. (2016).

They say in the Muslim religion that everybody, all our relatives that have died, watch over us, because when we pray and say prayers for them, they too do the same for us, so she [my deceased mother] continues to watch over us.

A Muslim young woman elaborated further:

Each time you pray, if you pray for the deceased, they will receive the prayers. They too pray, in return, so that you stay alive for a long time so that you can continue to pray for them.

Praying could also bring a welcome sense of presence of the deceased, as one Muslim young woman said: 'It's as if the person is next to you. They are a kind of memories.' This suggests that prayer affords not only an encounter with God, but also a means of becoming closer to a deceased relative and to the 'transcendent reality' of the dead more generally.

Prayers were also an important continuing practice of care for the deceased among Catholics, with some pointing to the deceased's presence in receiving their prayers. Simone, a Catholic widow (aged 39) commented: 'He receives my prayers (...). We don't see him, but he sees us.' As well as daily prayers and special Masses, many Catholics spoke of visiting the cemetery as a particular place for praying. However, amongst Muslims, it was very noticeable that it was only men who said that they visited cemeteries. Several Muslim interviewees mentioned reciting the Koran together regularly as a family as a way of caring for the deceased.

Some young people of both Muslim and Catholic religious affiliations saw continuing to pray and to care for other family members as practices which continued the wishes of the deceased. Albertine (aged 19, Catholic) said:

[I] pray every night before going to sleep because he [her father] was a man who believed in God, who loved his religion... I pray for him, for the house, for my brothers and sisters and everybody. I do that for him because he liked his children to pray and so on.

Similiarly, Magatte (aged 17, Muslim), said that she was learning the Koran to fulfil her deceased father's wishes. Religious practices thus offer a way to continue relationships with, and to please, the deceased.

An important practice of continuing bonds with the deceased, particularly for Muslims, was the giving of food or money to others as offerings or alms on significant religious days such as Fridays or during Ramadan or other religious festivals:

During the month of Ramadan, each day I prepare *ndogou* [what Muslims prepare for the breaking of the fast, usually coffee, milk or African herb tea, with bread and dates] and I give that to elderly people because my husband died during the month of Ramadan.

> (Athia, aged 56, widow, husband died three years ago)

Muslim interviewees said that they bought particular foods or prepared dishes that the deceased had enjoyed, which was given to others in the household or community (often children or older people) in remembrance of the dead. Others wished to give regular offerings on Fridays, but could not afford to.

Both Muslim and Catholic families gave accounts of religious ceremonies to remember their relative on the anniversary of the death or on a special religious occasion. Reciting the Koran was a central Muslim practice at such events. Several Catholics requested Mass to be said in church, when they could afford it and marked the anniversary of the death in particular ways. These included going to the cemetery with other family members to say prayers and light candles for their deceased relative, while others remembered the deceased particularly on All Saints Day. Many workshop participants said that religious ceremonies organised on the anniversary of the death were occasions when they or other family members could express their grief, as one Catholic young woman commented, 'When you organise a mass, it's inevitable you will cry'.

A common feature of continuing bonds is the sense of the presence of the deceased (Klass 2014). Several interviewees in Senegal acknowledged that they continued to feel their relative's presence, which seemed to be a source of consolation: 'I feel her presence from time to time. I was very close to her' (Allassane, man, aged 36, mother died two years previously). Similarly, a young woman whose mother had died ten years previously commented:

> I continue my life but I think of her when I sleep, when I walk. She is always beside me. She accompanies me in everything I do, so I feel her presence, despite the fact she has disappeared.

Steffen and Coyle's (2011) research with those of Christian and no religious affiliation in England found that the sense of presence was consoling to bereaved people when the continuing bond was understood within a framework of spiritual and religious meaningfulness. However, a conflict was often observed between their felt experience and culturally available explanatory frameworks. In our research, participants' responses about feeling the presence of their deceased relative were often rather brief, perhaps because this continuing spiritual connection to the deceased was difficult to reconcile with the dominant religious framing and cultural imperative to accept the death as 'God's will'. Nevertheless, one older widow expressed how her memories of shared moments with the deceased were ever-present and the impact of the loss did not subside over time:

I think each time you wake up and you remember she used to do that or he used to do this, that affects you all the time. It's as if it were today. Even if it's five years or something like that.

This powerfully conveys how the pain of grief and memories of the deceased are most often felt in the everyday mundane places and activities of a life shared together and may continue over the lifecourse, despite the solace often found in religious beliefs, practices and belonging to the wider family and community. This throws light on the earlier historical understanding of grief as being 'unconsoled' or '*ungetröstet*' in Freud's words, to which Klass (2014: 2) draws attention. It also highlights the importance to the living of their on-going relationships with the deceased.

Conclusion

This chapter has shown how participants in Senegal found consolation in *human relationships* characterised by non-judgemental co-presence, *cultural resources* provided by religious and cultural narratives and rituals, and *religious solace* by participating in something that transcends present space and time. Religious and cultural narratives about *Solidarité/ Dimbalanté* (solidarity/ mutual support), *Teranga* (hospitality), *Mounieul* (perseverence) and *Natu* (test of faith) provided the key frames of reference or 'worldview' (Klass 2014) that enabled bereaved family members in urban Senegal to make sense of their relative's death and be consoled, despite the pain of their loss. The frequent use of 'God's will' and other religious refrains and investing the moment of death with religious significance appeared to provide solace and help participants accept the death. An encounter with a transcendent reality at the moment of death, the religious worldview that gives meaning to the death and the faith community, all seemed to validate people's experiences.

The co-presence of family and community members in the immediate aftermath of the death was crucial in helping to share their pain, provide practical and material support and provide consolation that enabled family members to 'keep going' and 'get by' in poor urban neighbourhoods. The threat to social cohesion that a family death posed, in terms of the potential breaking up, and disrupting of, family relationships and resulting isolation, was thereby alleviated and bereaved family members were consoled that the reciprocal web of care would continue to support them.

For both Muslims and Catholics, prayers and religious ceremonies on the anniversary of the death were central practices of remembrance that enabled the embodied expression of continuing bonds with those who had died. The regular practice of prayer seemed to enable not only an encounter with the 'transcendent reality' of God (Klass 2014), but also brought people closer to their deceased relative, whom they often identified with the wider set of deceased family members, and with the dead more broadly. Regular offerings, in the form of giving food the deceased had liked to others on Fridays or during particular religious festivals, as well as holding recitals of the Koran, were particularly important to Muslims, while Catholics found solace in requesting a Mass or lighting candles in remembrance of the deceased. The living thus continued to feel connected to the dead through observing highly embodied practices on a regular (often daily, in the case of prayers) basis in everyday (non-sacred) spaces of the home and neighbourhood. Such practices of remembrance interweave religious and African cultural understandings and affirm the meanings the bereaved make of their relative's death and their relationship with the deceased, whilst also strengthening bereaved relatives' connections to reciprocal kinship and community support networks among the living.

We argue elsewhere (Ribbens McCarthy et al., in press) that responses to death can only be made sense of through an holistic approach, and in resource-constrained settings, such as the poor urban neighbourhoods where families participating in our research lived, the emotional and spiritual dimensions of loss and consolation were intrinsically interwoven with the material and social dimensions. While many participants spoke of the comfort their faith brought, and of their trust in God, which helped to resign themselves to what had happened, religious informal rules about the expression of grief also formed part of the social regulation of emotions. Crying too much or refusing to 'be strong' and 'keep going' might be viewed as a failure to accept the death, recognise God's will or to live up to the test of faith (*Natu*) that was involved. This demonstrates the extensive, often taken-for-granted, ways that religious beliefs, the worldview and the wider community shape cultural narratives of a family death and may limit the extent to which negative impacts on surviving family members' lives can sometimes be acknowledged in the Senegalese context (Evans et al. 2016). Thus, as Klass (2014: 15) observes, 'Consolation soothes and alleviates the burden of grief, but does not take away the pain'.

References

ANSD. 2013. *Rapport de la Deuxième Enquête de suivi de la pauvreté au Sénégal de 2011 (ESPS II)*. Dakar: Ministère de l'Economie et des Finances, République du Sénégal.

Bass, L. and Sow, F. 2006. Senegalese families: the confluence of ethnicity, history and social change, in *African Families at the Turn of the 21st Century*, edited by Y. Oheneba-Sakyi and B. Takyi. Westport: Praeger, 83–102.

Bignante, E. 2015. Therapeutic landscapes of traditional healing: building spaces of well-being with the traditional healer in St. Louis, Senegal. *Social and Cultural Geography* 16/6, 698–713.

Cacciatore, J. and De Frain, J., eds. 2015. *The World of Bereavement: Cultural Perspectives on Death in Families*. Cham: Springer.

Dunn, C., Le Mare, A., and Makungu, C. 2015. Connecting global health interventions and lived experiences: suspending 'normality' at funerals in rural Tanzania. *Social and Cultural Geography*, 17/2, 262–281.

Evans, R., Ribbens McCarthy, J., Kébé, F., Bowlby, S., and Wouango, J. 2017. Interpreting 'grief' in Senegal: language, emotions and cross-cultural translation in a francophone African context. *Mortality* 22/2, 118–135.

Evans, R., Ribbens McCarthy, J., Bowlby, S., Wouango, J., and Kébé, F. 2016. *Responses to Death, Care and Family Relations in Urban Senegal*, Research Report 1, Human Geography Research Cluster, University of Reading, Reading, UK. http://blogs.reading.ac.uk/deathinthefamilyinsenegal/report-key-findings/ [accessed 24/4/2018].

Falah, G.W. and Nagel, C. 2005. *Geographies of Muslim Women: Gender, Religion and Space*. New York: Guilford Press.

Gasparetti, F. 2011. Relying on 'Teranga': Senegalese migrants to Italy and their children left behind. *Autrepart* 1/57–58, 215–232.

Gökariksel, B. 2009. Beyond the officially sacred: religion, secularism, and the body in the production of subjectivity. *Social and Cultural Geography* 10/6, 657–674.

Hockey, J., Komovomy, C., and Woodthorpe, K. 2010. *The Matter of Death: Space, Place and Materiality*. Basingstoke: Palgrave.

Howarth, G. 2007. The rebirth of death: continuing relationships with the dead, in *Remember Me: Constructing Immortality. Beliefs on Immortality, Life, and Death*, edited by M. Mitchell. London: Routledge.

Jackson, M. 2004. The prose of suffering and the practice of silence. *Spiritus: A Journal of Christian Spirituality* 4, 44–59.

Jacquemin, M. 2010. *Urbanization, social change and child protection in West and Central Africa*. Dakar: UNICEF West and Central Africa Regional Office.

Klaits, F. 2005. The widow in blue: blood and morality of remembering in Botswana's time of AIDS. *Africa* 75/1, 46–62.

Klass, D. 1999. Developing a cross-cultural model of grief. *Omega* 39/3, 153–178.

Klass, D. 2014. Grief, consolation and religions: a conceptual framework. *Omega* 69/1, 1–18.

Klass, D., Silverman, P. R., and Nickman, S. L., eds. 1996. *Continuing Bonds: New Understandings of Grief*. Washington, DC: Taylor & Francis.

de Klerk, J. 2013. Being tough, being healthy: local forms of counselling in response to adult death in northwest Tanzania. *Culture, Health and Sexuality* 15/4, S482–S494.

Maddrell, A. 2009. A place for grief and belief: the Witness Cairn, Isle of Whithorn, Galloway, Scotland. *Social and Cultural Geography* 10/6, 675–693.

Maddrell, A. and Sidaway, J. 2010. *Deathscapes: Spaces for Death, Dying, Mourning and Remembrance*. Farnham: Ashgate.

Mills, A. and Gökariksel, B. 2014. Provincializing geographies of religion: Muslim identities beyond the 'west'. *Geography Compass* 8/12, 902–914.

Ndiaye, L. 2009. *Parenté et Mort chez les Wolof. Traditions et modernité au Sénégal*. Paris: L'Harmattan.

Prothmann, S. 2017. Ndogalu Yàlla: the judgement of God: migration aspirations and Sufi-Islam in urban Senegal. Paper presented at New Social Dynamics in Senegal workshop, Université Libre de Bruxelles, 13–14 March 2017.

Randall, S. and Coast, E. 2015. Poverty in African households: the limits of survey and census representations. *The Journal of Development Studies*, 51/2, 162–177.

Ribbens McCarthy, J., Evans, R., Bowlby, S., Wouango, J. and Kébé, F. (in press). Making sense of family deaths in urban Senegal: diversities, contexts and comparisons. *OMEGA – Journal of Death and Dying*.

Rosenblatt, P. and Bowman, T. 2013. Alternative approaches to conceptualizing grief: a conversation. *Bereavement Care* 32/2, 82–85.

Silverman, G. (2017). Kaddish and continuing bonds: an ethnographic exploration. Paper presented at workshop, Towards an Anthropology of Grief, 8th–9th of March 2017, Université Libre de Bruxelles, Brussels, Belgium.

Sogolo, G. 2003. The concept of cause in African thought, in *The African Philosophy Reader*, edited by P. H. Coetzee and A. P. J. Roux. Second edition. New York: Routledge, 192–199.

Sow, F. 2003. Fundamentalisms, globalisation and women's human rights in Senegal. *Gender and Development* 11/1, 69–76.

Steffen, E. and Coyle, A. 2011. Sense of presence experiences and meaning-making in bereavement: a qualitative analysis. *Death Studies* 35, 579–609.

Strange, J. M. 2005. *Death, Grief and Poverty in Britain, 1870–1914*. Cambridge: Cambridge University Press.

Valentine, C. 2008. *Bereavement Narratives: Continuing Bonds in the Twenty-First Century*. London: Routledge.

Walter, T. 1999. A death in our street. *Health and Place* 5, 119–124.

Wikan, U. 1988. Bereavement and Loss in two Muslim communities: Egypt and Bali compared. *Social Science and Medicine* 27/5, 451–460.

Conclusion

Analysing consolationscapes

Christoph Jedan

On two counts, the present collection's scholarly ambition is timely and challenging. First, consolation has fallen off the West's cultural radar, so that we are in danger of not 'getting' it. This neglect or misunderstanding of consolation extends right into today's academic scholarship. The philosopher Thomas Attig's excellent book *How We Grieve: Relearning the World* (2011) is a case in point. Building on Colin Murray Parkes' ideas that grieving involves a loss of the assumptive world and that the bereaved need to 'relearn the world', Attig's book demonstrates with a rich phenomenology that adjusting to loss is far more than an abstract cognitive process; it involves every facet of our being-in-the-world. Yet consolation plays only a marginal role in that book. In the few passing references, consolation is regularly married to religious beliefs, often with a ring of insincerity and ineffectiveness. The story of the death of six-year-old Bobby is a case in point: 'The funeral seems but a necessary formality. A minister attempts to offer peace and consolation through words that Ed and Elise hear as but a string of meaningless clichés and platitudes' (2011: 101–102). If we were to follow such interpretative templates, consolation might appear to be a 'toxic brand' in the cultural situation of the early twenty-first century. The neglect of consolation has been rightly highlighted by Dennis Klass. 'Consolation,' he writes, 'is grief's traditional amelioration, but contemporary bereavement theory lacks a conceptual framework to include it' (Klass, unpublished). Klass himself suggests a highly valuable tripartite framework to fill the lacuna (Klass 1993; 2006; 2014), and I regard my own Four-Axis Model, presented in Chapter 1, as a complementary attempt to bring to such frameworks a stronger emphasis on the historical experience than has been brought to date.

Second, many of the humanities disciplines have witnessed a veritable 'spatial turn', so that more researchers than ever before recognise spatial constellations as key to the phenomena they describe. Once again, the death of six-year-old Bobby as related in Thomas Attig's *How We Grieve* is a good example. Attig describes how Bobby's death haunts his father Ed 'in every corner of the house' (2011: 102), rendering certain rooms too painful to enter, and the sight of Bobby's toys too upsetting for them to be left lying around. The entire case-study abounds with spatial vocabulary:

When Ed and Elise return home first from the hospital where Bobby has died and later from Bobby's funeral and burial, they face a world that is changed utterly by what has happened. They can never experience, or be at home in, that world in the same way they were prior to Bobby's death. They are reminded of Bobby's absence everywhere, by the things he has left behind, in the places where they shared life with him, in interaction with one another and with others who survive with them, and in their own minds and hearts where they came to know and love him. Relearning their worlds is not simply a matter of registering Bobby's absence or taking in new information about the world as it is now without him. It is a struggle to discover, and make their own, ways of going on without him in that world.

(Attig 2011: 105)

The concept of 'relearning the world' obviously has spatial aspects and lends itself to analysis from a spatial perspective. However, researchers from the humanities have been slow to extrapolate higher-level spatial frameworks of bereavement from their idiographic descriptions. Avril Maddrell's framework of three grief/consolation spaces brings the much-needed systematicity to idiographic descriptions. Her model is not the only spatial framework around, but – as I can testify from my attempts to apply it in Chapter 1 – it is an eminently useful one.

In short, what is needed today are analyses of *consolationscapes*, i.e. analyses of the many ways in which consolation and spatial constellations are intertwined and historically inflected. This is exactly what Avril Maddrell, Eric Venbrux and I have envisaged with the present volume. Analyses of consolationscapes will not only show with concrete case studies how consolation and space intersect; they will also demonstrate by their results that consolation is still a fertile concept for analysing human responses to losses, to those past as well as to those present.

As editors, Avril Maddrell, Eric Venbrux and I have taken care not to limit the discussions in this volume. The authors were not asked to relate to and/or comment on the conceptual frameworks proposed by Avril Maddrell and me. While we are convinced that the frameworks will prove useful for future research, we felt it important not to foreclose other perspectives in a volume that attempts to break fresh ground. This opens up delightful possibilities of conversation about the theoretical frameworks contained in the first section of this book, and about the case-studies in its second and third sections. It is tempting to speculate what new questions might be generated by the conceptual frameworks proposed in the volume, and how emphases in the specific case-studies might change on their base.

In Chapter 10, for instance, Ruth Evans, Sophie Bowlby, Jane Ribbens McCarthy, Joséphine Wouango and Fatou Kébé put Klass's tripartite framework to excellent use. In so doing, they highlight 'religious' formulas such as 'It's God's will', which could be analysed with the Four-Axis Model as invoking a 'healing' world-view and view of death in particular (death as under God's control, the deceased being looked after by God). Axis 3 of the Four-Axis Model, however, does not single out any religious motifs. The question suggested by a comparison between that model and Chapter 10 is this: Are there any other, not so overtly 'religious' world-view elements invoked by the interviewees? If there are none or if their presence is not very conspicuous, how can the researchers account for their absence? Comparison between the Four-Axis Model and Chapter 10 also suggests other interesting questions: What are the ideals of acceptable grief informing the interviewees' answers? Are there explicit descriptions of such an ideal in their answers (Axis 1)? Do we find appeals to resilience in the interviewees' thick descriptions of consolatory practices (Axis 2)? And, whilst memorialisation is highlighted in Chapter 10 as part of the practice of maintaining continuing bonds, Axis 4 suggests the importance of *inter alia* virtues, the uniqueness of the individual, and the completion of landmark tasks or bucket-lists as historically important ways of preserving the wholeness of the deceased's life. Are these ways represented in the interview material and, if so, how?

A similar conversation could be staged over Avril Maddrell's tripartite framework and the anthropological chapters in the book's third section. Her tripartite model distinguishes physical or material spaces from embodied-psychological and virtual spaces. Comparison shows that some of the chapters can be analysed as prioritising one or other of her categories. To take again Chapter 10 as an example, it is clear that by taking Klass's model as the point of departure the chapter's main emphasis lies on embodied-psychological spaces; by contrast, physical or material spaces – how they are perceived, demarcated and preserved by ritual, and so on – receive some, but arguably less, attention. The converse holds for Eric Venbrux's Chapter 7, where physical or material spaces and their role in ritual take centre stage. None of the chapters in the third section, however, explores in any detail Avril Maddrell's third category of virtual space. To me, this seems an important avenue for future research on consolationscapes: there is emerging interest in anthropologies of technology beyond the Global North (see Telban and Vávrová 2014), and researchers into death, grief and resilience should follow this line of enquiry lest they perpetuate as low-technology the exoticising depictions of the Global South.

Of course, such a conversation would not be one-sided. From the perspective of the case studies one could reply, for instance, that the Four-Axis Model emphasises cognitive and intellectual aspects of consolation. The model is unabashedly world-view-centric, and whilst cognitive and world-view aspects are important in ritual and in formulaic behaviour, one should not underrate the comforting function of repetition, sometimes even at the expense of intellectual structure. Moreover, even when accepting the distinction between three categories of grief and consolation spaces as a useful conceptual repertoire, one might maintain that in different situations the relative importance of these categories will vary.

I envisage, in short, an open-ended conversation that would be likely to lead to the further adjustment of idiographic case studies as well as to a further refinement of the conceptual frameworks. There is no way of knowing exactly how this conversation might end. All I can suggest is that it is important for us to engage with it in our cultural circumstances today and to try to involve other approaches such as psychologies of grief and bereavement – approaches that not only stand to gain from the concept of consolationscapes, but are also likely to bring fresh insights to the debate.

References

Attig, T. 2011. *How We Grieve: Relearning the World*. Second edition. New York: Oxford University Press.

Klass, D. 1993. Solace and immortality: bereaved parents' continuing bonds with their children. *Death Studies* 17/4, 343–368.

Klass, D. 2006. Grief, religion, and spirituality, in *Death and Religion in a Changing World*, edited by K. Garces-Foley. Armonk, NY: Sharpe, 283–304.

Klass, D. 2014. Grief, consolation, and religions: a conceptual framework. *Omega* 69/1, 1–18.

Klass, D. (unpublished). The nature of religious consolation for the bereaved. Unpublished draft. Available at: http://www.academia.edu/2247617/The_Nature_of_Religious_Consolation_for_the_Bereaved.

Telban, B. and Vávrová, D. 2014. Ringing the living and the dead: mobile phones in a Sepik society. *The Australian Journal of Anthropology* 25, 223–238.

Index